WiMax Operator's Manual

Building 802.16 Wireless Networks (Second Edition)

■ ■ ■

Daniel Sweeney

Apress®

WiMax Operator's Manual: Building 802.16 Wireless Networks (Second Edition)

ISBN (pbk): 1-59059-574-2

Printed and bound in the United States of America 9 8 7 6 5 4 3 2 1

Lead Editor: Jim Sumser
Technical Reviewer: Robert Hoskins
Editorial Board: Steve Anglin, Dan Appleman, Ewan Buckingham, Gary Cornell, Tony Davis, Jason Gilmore, Jonathan Hassell, Chris Mills, Dominic Shakeshaft, Jim Sumser
Project Manager: Denise Santoro Lincoln
Copy Edit Manager: Nicole LeClerc
Copy Editor: Liz Welch
Assistant Production Director: Kari Brooks-Copony
Production Editor: Kari Brooks-Copony
Compositor: Pat Christenson
Proofreader: Lori Bring
Indexer: Valerie Perry
Cover Designer: Kurt Krames
Manufacturing Director: Tom Debolski

Distributed to the book trade worldwide by Springer-Verlag New York, Inc., 233 Spring Street, 6th Floor, New York, NY 10013. Phone 1-800-SPRINGER, fax 201-348-4505, e-mail orders-ny@springer-sbm.com, or visit http://www.springeronline.com.

For information on translations, please contact Apress directly at 2560 Ninth Street, Suite 219, Berkeley, CA 94710. Phone 510-549-5930, fax 510-549-5939, e-mail info@apress.com, or visit http://www.apress.com.

This book is dedicated to my wife.

Contents at a Glance

Contents

About the Author

DANIEL SWEENEY is a technical writer, business reporter, and industry analyst. He has written thousands of articles and several analyst reports. He covers telecommunications, consumer electronics, energy, and the history of technology, with occasional forays into military technology, artificial intelligence, and geology. He has written for leading trade journals in telecommunications and both trade and consumer journals in consumer electronics. In the past he worked as a common laborer, a labor organizer, and a government bureaucrat who compiled mind-numbing statistical reports. He is married and lives in the vicinity of a toxic waste dump (seriously).

About the Technical Reviewer

ROBERT HOSKINS is the publisher and editor of Broadband Wireless Exchange, the leading online publication in the field, and is a former Sprint executive responsible for managing what is still the largest and most successful broadband wireless deployment in the United States.

Preface

The second edition of *WiMax Operator's Manual* includes most of the material from the first edition, plus new discussions of

- The ultra-high-speed mobile telephone standard, HSDPA
- Ultrawideband (UWB)
- Changes to DSL technologies
- Mobile voice
- Mobile entertainment
- New backup systems

The new edition also reflects the changes that have occurred in the industry over the last year and half, including the emergence of prestandards wireless broadband equipment with fully developed mobile capabilities, significant alterations in the competitive landscape, and the opening of valuable new spectrum for broadband wireless operators.

Public broadband wireless data networks represent a truly disruptive technology, one that promises to break the monopolistic and oligopolistic status quo that still represents the norm in high-speed access today. Products that would enable such networks have existed for a number of years and in fact have been deployed in thousands of commercial systems throughout the world, but the lack of standards, the limited production volumes, and the consequent high prices have prevented the full potential of wireless broadband from being realized. Now, with the coming of a widely accepted industry standard, IEEE 802.16, and the introduction of microchips based on that standard by leading semiconductor companies, wireless broadband public networks are becoming mainstream.

Working as a journalist, analyst, and consultant in the field of telecommunications, I have been covering wireless broadband extensively since 1990, before public networks even emerged, and I've witnessed the steady progress of the technology as well as the many false starts of the wireless broadband industry. And for the first time I can report with some confidence that wireless broadband is ready to compete in the marketplace.

As in the past, wireless will continue to attract entrepreneurs—in many cases, entrepreneurs lacking in experience in either telecommunications or radio frequency electronics. Such individuals will face a "steep learning curve" and will have to acquire working knowledge in both areas in order to stand a chance of succeeding. It is my hope that this book, based on dozens of case histories and my own considerable experience in both fields, will provide such individuals with a wireless broadband survival kit.

Introduction

Broadband wireless has long held the promise of delivering a wide range of data and information services to business and residential customers quickly and cost-effectively. Unfortunately, that promise has been imperfectly met in the past because of both the immaturity of the existing technologies and the relatively high cost of networking equipment. With the publication of a comprehensive industry standard—namely, IEEE 802.16—representing a distillation of the most advanced technology and an industry consensus permitting equipment interoperability, broadband wireless has gained the maturity it lacked and is truly ready for utilization within metropolitan networks.

The first chips adhering in full to the 802.16 standard have begun to be shipped by a number of semiconductor manufacturers. Some time will pass before such chips appear in assembled systems and before they are certified for standards compliance and interoperability, but even now broadband wireless can be said to be approaching the stage of early maturity. Such developments will provide the basis for broadband wireless establishing a real competitive presence in the marketplace, something it has never enjoyed in the past.

This book provides the background in broadband wireless fundamentals, packet data, and overall network operation and management to enable a network operator to set up a network with standards-based equipment and to run it profitably thereafter. It is an operational handbook rather than an engineering text, and it is highly practical rather than theoretical. Technical discussions that occur are always in reference to addressing the real-world problems involved in running a network and serving the customer base. There are no tutorials on radio frequency propagation or digital modulation techniques; rather, the emphasis is on using technology to deliver specific services to specific types of customers.

Broadband wireless as a last-mile access technology is a fairly recent phenomenon, and most of the success stories are recent. Not a lot of standard procedures are extant in the marketplace for operating a network successfully, and not a lot of network executives and managers have a deep knowledge of broadband wireless. And scarcely any texts at all provide compendia of facts and analysis on the subject. This book meets a real need for a concise summary source of information.

Broadband wireless at this point still represents a divergent, even disruptive, technology, and wireline solutions such as fiber optics, hybrid fiber coax, and digital subscriber line (DSL) constitute the mainstream. For this reason, a great many of broadband wireless ventures to date have been highly speculative and entrepreneurial, with many of the pioneers painfully attempting to find their way even as their networks were in the process of being built. This book serves as a guide for present and future entrepreneurs and is intended to assist them in avoiding the experiments and false starts that proved so frustrating for the pioneers.

Since this book is utilitarian rather than highly conceptual, it does not constitute the sum of all information relating to broadband wireless networks. What this text contains is a body of immediately practical knowledge—what to do and how to do it. And perhaps most important, it explains who the appropriate professionals and technicians are to retain when initiating and maintaining a broadband wireless network. The book presents such knowledge from a business perspective with a just consideration of likely costs and payoffs and with the caveat that almost any decision made in regard to the network is provisional and ultimately dependent on the changing nature of the customer base, the regulatory environment, the financial markets, the competitive atmosphere, and of course ongoing advances in technology.

Tremendous strides have been made in digital radio technology over the course of the last decade, culminating in the 802.16 standard, and wireless has emerged as viable broadband access technology where it was marginal at best as recently as four years ago. In many instances, wireless broadband is the preferred access technology, offering the best cost/performance ratio, time to market, and service velocity. Still, it does not always enjoy a competitive advantage, and in many markets a broadband wireless solution may be suboptimal or even ill advised. The physical layer is but one part of the service network, and insisting on wireless for its own sake while ignoring overall network architecture makes little sense. The physical layer, the access layer, and all the intervening layers ultimately support the topmost layer (namely, applications), and the issue that must always be uppermost in the mind of the network operator is how the applications and services wanted by subscribers can be delivered most cost effectively. If the answer includes a wireless physical link, then a complete perusal of the contents of this book is indicated. If the answer is otherwise, then Chapters 1 and 2 will provide all of the information one needs.

Finally, it should be understood that wireless can and often is used in piecemeal fashion to extend wireline infrastructure, and following such a course is not at all illegitimate or even ad hoc. Nothing is particularly admirable about purism in terms of wireless technology, and if wireline technologies serve the same purpose better over some portion of the network footprint, then wise network operators will avail themselves of them.

Unfortunately, no department of broadband wireless administration exists in any university of which I am aware. Such knowledge as I have obtained has been from various scattered engineering texts and from those individuals who have developed the products and procedures and have overseen the implementation of the first successful networks. Their names are legion, and I cannot thank all of them, but I will mention the following individuals who have taught me much: Bill Frezza of Adams Venture Capital, Craig Matthias of the FarPoint Group, Doug Lockie of Endwave Corporation, and Keith Reid of Cisco Systems. Any inaccuracies in this text must be laid to my account and not to any of them.

CHAPTER 1

Wireless Broadband and the Standards Governing It

This book focuses on standards-based public broadband wireless networks operated on a per-profit basis. In the past many broadband wireless networks utilizing equipment not based on standards or utilizing equipment based on wireless local area network (WLAN) standards have been launched only to fail within a short period. The emergence of standards-based equipment stemming from the specific needs of the public service provider marks a momentous change in the broadband marketplace and will enable wireless networks to take their place beside successful wireline services such as optical fiber networks, digital subscriber line (DSL), and cable. The appearance of such equipment will also enable the network operator to generate consistent revenues and to attract and retain valued customers, provided, that is, that the operator understands both the strengths and the limitations of the technology and comprehends how to run a network in a businesslike manner.

Defining Wireless Broadband

The term *wireless broadband* generally refers to high-speed (minimally, several hundred kilobits per second) data transmissions occurring within an infrastructure of more or less fixed points, including both stationary subscriber terminals and service provider base stations (which themselves constitute the hubs of the network). This is distinct from mobile data transmissions where the subscriber can expect to access the network while in transit and where only the network operator's base stations occupy fixed locations. You can expect that this distinction will become somewhat blurred in the future inasmuch as several manufacturers are developing very high-speed wireless networking equipment that will support mobility or stationary usage almost equally well, but the emphasis of high-speed wireless service providers serving stationary subscribers will remain. Broadband wireless, as it is today, is properly a competitor to optical fiber, hybrid fiber coax (the physical infrastructure of most cable television plants), DSL, and, to a much lesser extent, broadband satellite.

Third-generation (3G) and 2.5G cellular telephone networks, which have special provisions for delivering medium-speed packet data services, have not, in most instances, been directly competitive with broadband wireless services. They share a radio frequency airlink and, in some cases, core technologies, but they have traditionally served a different type of customer and have presented different types of service offerings.

This may be changing. Recently, a new mobile standard known as High-Speed Downlink Packet Access (HSDPA) has emerged, and the first networks utilizing it are already in operation in Asia. HSDPA, which is an extension of Global System for Mobile Communications (GSM), the most widely used standard for digital cellular telephony, supports throughputs exceeding 10 megabits per second (Mbps) while affording full mobility to the user. An HSDPA capability, which may easily and inexpensively be added to an existing GSM network, provides the network operator with a true broadband service offering capable of competing with cable or DSL data services. GSM networks, for the most part, still face the challenge imposed by bandwidth allocations that are marginal for provisioning large numbers of broadband customers, but HSDPA definitely undercuts many of the assumptions in the marketplace on the limitations of mobile services and appears to pose a real alternative.

Whether that alternative will be sufficient to retard the acceptance of 802.16 in the broadband marketplace remains to be determined. HSDPA will be utilized almost exclusively by existing mobile license holders, in most cases large incumbents with multiple local networks extending over a national footprint. 802.16, on the other hand, is likely to be the province of independents or of non-telco wireline operators such as cable networks that are seeking a wireless and, in many cases, a mobile offering. Because of the differences in service orientation that characterize the two camps, the service bundles actually offered to the public are likely to be different and the outcome of the contest between HSDPA and 802.16 will probably depend as much on market positioning as on the capacities of either technology. At the same time, the fact that the mobile operators possess built-out physical infrastructure and can leverage it effectively to deploy HSDPA either rapidly or incrementally, depending on their strategies, means that challengers operating 802.16 networks will face formidable opposition in the markets where HSDPA gains a foothold.

Introducing the 802.16 Standard

A number of industry standards govern the design and performance of wireless broadband equipment. The standards that chiefly concern wireless broadband are 802.16 and its derivative 802.16a, both of which were developed by the Institute of Electrical and Electronic Engineers (IEEE), a major industry standards body headquartered in the United States.

The complete standards are available as book-length documents on the IEEE Web site at `http://www.ieee.org`. This chapter focuses on only the most salient points in respect to network operators.

Both standards have as their goal the standardization of acceptable performance levels and the achievement of full interoperability among the products of standards-compliant manufacturers. The latter will allow the network operators to mix base stations and subscriber premises equipment from different manufacturers so as not to be dependent on single sourcing and, perhaps more important, to encourage the mass production of standards-based chipsets by competing manufacturers. This in turn will lead to a drop in equipment prices because of economies of scale and market pressures.

In the past, the high prices of carrier-grade wireless base stations and subscriber terminals have saddled network operators with unacceptable equipment costs, and such costs, coupled with the disappointing performance of first-generation products, severely hindered wireless network operators attempting to compete with wireline operators. The present availability of substantially better-performing and less-expensive infrastructure equipment should finally enable network operators to utilize wireless access technologies advantageously and compete effectively with wireline broadband services.

The 802.16 and 802.16a standards share the same media access control (MAC) layer specifications but posit different physical layers because of the different areas of spectrum covered by the respective standards. The 802.16 standard covers what has come to be known as the *millimeter microwave spectrum* and extends from 10 gigahertz (GHz) up to 66GHz, and 802.16a covers 2GHz to 11GHz; the two standards thus overlap. In fact, most of the activity involving 802.16a-based equipment is likely to occur at frequencies below 6GHz because lower-microwave equipment is both less expensive and more versatile.

Unlike the standards governing WLANs (namely, 802.11 and its derivatives—802.11b, 802.11a, 802.11g, 802.11e, 802.11n, 802.11p, and 802.11s), the 802.16 standards do not state fixed throughput rates for the individual user but state only a maximum of 124Mbps for a channel for 802.16 and 70Mbps for a 20 megahertz (MHz) channel bandwidth in the 802.16a standard. In fact, the lack of stated rates is entirely appropriate to a standard intended for a public service provider because the operator needs to have the flexibility of assigning spectrum selectively and preferentially and of giving customers willing to pay for such services high continuous bit rates at the expense of lower-tier users—and conversely throttling bandwidth to such lower-tier users in the event of network congestion. In a public network, the operator and not the standard should set bit rates such that the bit rates are based on business decisions rather than artificial limits imposed by the protocol.

Introducing the Media Access Control Layer

The *media access control layer* refers to the network layer immediately above the physical layer, which is the actual physical medium for conveying data. The access layer, as the name suggests, determines the way in which subscribers access the network and how network resources are assigned to them.

The media access control layer described in the 802.16 standard is designed primarily to support point-to-multipoint (PTMP) network architectures, though it also supports the point-to-point (PTP) and point-to-consecutive point (PTCP) architectures. The lower-frequency bands also support mesh topologies, although the mesh standard adopted by the 802.11 committee does not reflect the latest research into mesh networking. Chapter 3 fully explains these terms.

The 802.16 standard has been optimized for Internet Protocol (IP) traffic, and IP-based services represent the best approach for most operators; however, standards-based equipment will also support legacy circuit-based services such as T1/E1 and asynchronous transfer mode (ATM). In general, the older circuit-based services represent inefficient use of bandwidth, an important consideration with wireless where bandwidth is usually at a premium. Moreover, they put the wireless broadband operator in the position of having to compete directly with the incumbent wireline telephone operator. Wireless insurgents attempting to vie for circuit traffic with strong, entrenched incumbents have been almost uniformly unsuccessful for reasons Chapter 6 will fully explore.

A few words about the circuit and quasi-circuit protocols: A *circuit transmission* is one in which a prescribed amount of bandwidth is reserved and made available to a single user exclusively for the duration of the transmission; in other words, the user occupies an individual channel. In a *packet transmission,* a channel is shared among a number of users, with each user transmitting bursts of data as traffic permits.

The T1/E1 terms mentioned previously refer to two closely related standard circuit-based service offerings delivered over aggregations of ordinary copper telephone wires. A T1, the American standard, consists of 24 copper pairs, each capable of a throughput speed of 64 kilobits per second (Kbps). E1 consists of 30 pairs and is commensurately faster. E1 is the standard offering in most countries outside the United States. A T1 is delivered over a synchronous optical network (SONET), which is covered in the following chapters. An E1 is delivered over a synchronous digital hierarchy (SDH) network, the European equivalent of SONET. Both services go through ordinary telephone switches to reach the subscriber.

ATM is a predominantly layer-2 (the switching layer) protocol developed in large part by Bellcore, the research arm of the Bell Operating Companies in the United States. Intended to provide a common platform for voice, data, and multimedia that would surpass the efficiency of traditional circuit networks while providing bandwidth reservation and quality-of-service (QoS) mechanisms that emulate circuit predictability, ATM has found its place at the core of long-haul networks where its traffic-shaping capabilities have proven particularly useful. In metropolitan area networks it is chiefly used for the transportation of frame-relay fast-packet business services and for the aggregation of DSL traffic. The 802.16 standard obviates the need for ATM, however, by providing comparable mechanisms of its own for bandwidth reservation and service-level stratification. Because ATM switches are extremely expensive and represent legacy technology, I do not recommend using ATM as a basis for the service network, unless, of course, the wireless network is an extension of an existing wired network anchored with ATM switches.

The 802.16 standard can accommodate both continuous and bursty traffic, but it uses what is essentially a connection-oriented protocol somewhat akin to those of ATM and frame relay. Modulation and coding schemes may be adjusted individually for each subscriber and may be dynamically adjusted during the course of a transmission to cope with the changing radio frequency (RF) environment. In the higher frequencies, 16 quadrature amplitude modulation (QUAM) and 64 QUAM are automatically invoked by the protocol to match signal characteristics with network conditions, with 64 QUAM providing greater information density and 16 QUAM providing greater robustness. The orthogonal frequency division multiplexing (OFDM) modulation scheme is specified for the lower band with a single carrier option being provided as well. Chapter 4 discusses these terms.

The 802.16 protocols are highly adaptive, and they enable subscriber terminals to signal their needs while at the same time allowing the base station to adjust operating parameters and power levels to meet subscriber needs. Polling on the part of the subscriber station is generally utilized to initiate a session, avoiding the simple contention-based network access schemes utilized for WLANs, but the network operator also has the option of assigning permanent virtual circuits to subscribers—essentially reservations of bandwidth. Provisions for privacy, security, and authentication of subscribers also exist. Advanced network management capabilities extending to layer 2 and above are not included in the standard.

Introducing the Two Physical Standards

The 802.16 standard requires two separate physical-layer standards because the propagation characteristics of radio waves are so different in the lower- and upper-microwave regions.

Lower-frequency signals can penetrate walls and can travel over considerable distances—more than 30 miles with highly directional antennas. Higher-frequency transmissions, on the other hand, must meet strict line-of-sight requirements and are usually restricted to distances of a few kilometers. The lower-frequency ranges also lend themselves to complex modulation techniques such as OFDM and wideband Code-Division Multiple Access (CDMA). These conduce to high levels of robustness and higher spectral efficiencies—that is, more users per a given allocation of bandwidth.

The singular advantage enjoyed by users of higher-frequency bands is an abundance of bandwidth. Most spectral assignments above 20GHz provide for several hundred megahertz minimally, and the 57GHz to 64GHz unlicensed band available in the United States can support several gigabits per second at one bit per hertz for fiberlike speeds.

Introducing WiMAX

Standards are of relatively little value unless there is some way of enforcing compliance to the standard. Promoters of 802.16 were well aware of this, and some of them elected to form an organization to test and certify products for interoperability and standards compliance. That organization is known as the Worldwide Interoperability for Microwave Access (WiMAX).

WiMAX also promotes the 802.16 standard and the development of what it calls *systems profiles*. These are specific implementations, selections of options within the standard, to suit particular ensembles of service offerings and subscriber populations.

At the time of this writing, the WiMAX has not certified any equipment designed according to the 802.16 standards, although the first 802.16 chips have reached the market and some have been submitted to the organization for evaluation and testing. WiMAX itself expects that some products will be certified by the end of 2005, but this is only an estimate. For this reason, the 802.16 network equipment that the operator intends on using today cannot be assumed to provide total interoperability.

Currently 802.16 chips are being shipped or have been announced by Intel, Fujitsu, Wavesat, Sequans, TeleCIS, Beceem Communications, Adaptix, and picoChip. WiMAX certification of at least some of these products will follow in 2006. Most industry observers believe that incorporation of first-generation chips in products will take place on a fairly small scale and that radio manufacturers are awaiting the finalization of the 802.16e mobility standard before committing to volume production.

Introducing Other Wireless Broadband Standards

An earlier IEEE standard, 802.11, and its derivatives (802.11b, 802.11a, 802.11g, and soon 802.11e) have seen wide deployment in commercial, governmental, and residential LAN settings and some application in public service networks, primarily localized *hotspots* where coverage is provided within a picocell not exceeding a couple of hundred yards in radius. I anticipate that low-priced 802.11 solutions will continue to be attempted within pervasive metropolitan networks better served by 802.16-based equipment. Table 1-1 compares the two standards in detail.

Table 1-1. *Standards: 802.11 vs. 802.16*

Feature	802.11	802.11b	802.11a	802.11g	802.16	802.16a
Assigned spectrum	2.46GHz	2.4GHz	5.8GHz	2.4GHz	10GHz–66GHz	2GHz–11GHz
Maximum throughput	2Mbps	11Mbps	54Mbps	54Mbps		70Mbps
Propagation distance	200 yards	200 yards	200 yards	200 yards	More than a mile	Several miles
Network architecture supported	Point to multipoint PTMP	PTMP	PTMP	PTMP	PTP, PTCM	PTMP, PTCM, mesh
Transport protocols supported	Ethernet	Ethernet	Ethernet	Ethernet	TCP, ATM	TCP/IP, ATM
Modulation system	Frequency hopping, direct sequence	Frequency hopping, direct sequence	OFDM	OFDM	QUAM, phase shift keying (PSK)	OFDM
Adaptive modulation?	No	No	No	No	Yes	Yes
Support for full mobility?	No	No	No	No	No	Upcoming
QoS?	No	No	No	No	Yes	Yes

I will not devote much attention in this book to the specifics of the 802.11 standard. Optimized for indoor and campus environments, 802.11 was intended to serve the needs of Ethernet LAN users and is quite limited in terms of range and the number of users that can be accommodated simultaneously. In fact, transmission speed and signal integrity drop off precipitately at distances beyond about 500 feet from an access point.

So why, given the intentional limitations inherent in the standard, would anyone contemplate employing 802.11 equipment in a public setting? In a word, price. Specifically, 802.11 gear has become a commodity; also, network interface cards for subscriber terminals are available at the time of this writing for less than $100, and access points are available for less than $200. Simply put, a network constructed of 802.11 network elements will cost a fraction of the amount of money required to purchase 802.16 equipment.

If the network consists of nothing but short cell radius hotspots, 802.11 will suffice and indeed may be preferable, but for a metropolitan network most 802.11 equipment represents a severe compromise. A few manufacturers (such as Tropos, Vivato, and Airgo) are attempting to manufacture adaptive-array antenna systems or mesh-networked base stations for 802.11-compliant equipment, expedients that will presumably emulate some of the characteristics of 802.16 in respect to distance and reuse of spectrum within a cell, but such equipment is much more expensive than conventional 802.11 products and still lacks the full complement of protocols for supporting QoS and advanced network management. Possibly in some situations, such “hotrodded” 802.11 gear will be adequate and will represent the most cost-effective equipment choice, but to regard it as a general substitute for 802.16 infrastructure is misguided. When a service provider is attempting to serve a number of small businesses in a single building where the subscribers lack networking capabilities of their own, enhanced performance 802.11 may be adequate, provided that the base station assigned to the building does not have to serve a larger area. No one should be tempted to believe that an entire metropolitan market can be served with 802.11 equipment.

At the same time, 802.11 in its various iterations definitely bears watching. The standard has been subject to continuous revision ever since it was introduced in 1998, and it has definitely not solidified as yet. Further revisions of the standard are in preparation, and some of these could render further generations of 802.11 equipment more competitive with 802.16 in the context of the metropolitan area network (MAN).

Of particular interest is the proposed 802.11e revision, which is currently in committee. The standard endows 802.11 networks with QoS capabilities. The standard calls for prioritization of different traffic types and also allows for contention-free transmissions to take place over short durations of time, a provision that would significantly reduce latency. But because 802.11 remains Ethernet rather than IP based, as is the case with 802.16, a comparable range of ancillary QoS protocols is not available. 802.11e may be entirely sufficient for transmitting time-sensitive traffic such as voice or video within a LAN environment, but its ability to maintain QoS across the metro may be questioned. Rumor has it that the IEEE will ratify 802.11e some time in 2006. Its appearance in silicon would probably take place a year or so later.

Also of considerable interest is 802.11n, the high-speed standard. Projected speeds are in excess of 100Mbps. Two variants are currently in contention: the World-Wide Spectrum Efficiency (WWiSe) specification backed by Broadcom and TGn Sync, supported by Intel and Philips, among others. Intel, it should be noted, has not previously been a player in the wireless fidelity (WiFi) space, and by devising a new 802.11 standard it would be redressing past deficiencies. The new standard will definitely make use of multiple input, multiple output (MIMO) technology, where arrays of antennas are required for both base stations and subscriber terminals. Ratification is expected to take place in late 2006. Incidentally, many manufacturers are discussing a standard beyond 802.11n that has yet to gain a number designation and is simply known as Gigabit 802.11. Achieving such throughputs over any unlicensed band currently accessible to 802.11 radios would be a major challenge with existing technology, and I think gigabit throughputs are still years away.

Finally, a mesh standard named 802.11s is in preparation. What effect this will have on the positioning of 802.11 vis-à-vis 802.16 is difficult to determine at present.

The 802.15 standard is another IEEE fixed-point wireless broadband standard, but it is one of even less relevance to public networks. The 802.15 standard incorporates an older standard promoted and largely developed by Ericsson and Nokia known as Bluetooth (named after Harald Bluetooth, a tenth-century Viking monarch). Bluetooth has been used in a few hotspot public networks, but the range is so short—no more than 50 yards or so—that it is utterly inapplicable in pervasive MANs. Also, work is under way on the formulation of a substandard within 802.15 to include ultrawideband (UWB) radio, a revolutionary RF technology that uses extremely low-power, wideband pulses intended to coexist with channelized radio communications. UWB may well represent the far future of broadband wireless, but current power restrictions confine it to very short ranges, just as with Bluetooth, and it is not suitable for overarching MANs as it is currently configured.

Finally, I should briefly mention High-Performance Radio Metropolitan Area Network (HIPERMAN), a standard that is somewhat analogous to 802.16 but that emanates from a different standards body, namely the European Telecommunications Standards Institute

(ETSI). HIPERMAN is an outgrowth of an earlier stillborn standard called HIPERLAN2. HIPERLAN2 appears to have been positioned to play in two distinct markets, one encompassing large, campuswide corporate LANs and one for commercial MANs. In some respects HIPERLAN2 is reminiscent of IEEE 802.11a, and in others it is closer to IEEE 802.16a. HIPERLAN2—which, as the name suggests, proceeded from HIPERLAN, an abortive standard that saw embodiment in only a few products made by Proxim—appears to have found even less market success than its predecessor, which is to say precisely none to date. The marketplace has already decisively rejected HIPERLAN in all its forms, but HIPERMAN is still being actively promoted by ETSI. Currently discussions are under way between the IEEE and ETSI toward merging HIPERMAN with 802.16. As it stands, no equipment embodying the HIPERMAN standard is in the marketplace.

Deploying Within Urban, Suburban, and Rural Environments

The IEEE 802.16 standards represent the institutionalization of several of the best-performing technologies in wireless communications and the aggregation of a number of advances made by various manufacturers that are unavailable in a single platform up to this time. As such, the new standards-based equipment enables broadband wireless networks to perform at a level that was unattainable previously and extends the capabilities of wireless access technologies to permit the penetration of markets where previously wireless broadband was marginal or simply ineffective.

Broadband wireless is still not the best access technology for all geographical markets or all market segments within a given geography, but many more customers are potentially accessible than in the past. It is scarcely an exaggeration to say that the new standards provide an effective solution to the most severe geographical limitations of traditional broadband wireless products, though the reach of any given wireless network is still constrained by its location, and its attractiveness is affected by the presence or absence of competing broadband technologies.

The most difficult geographical markets for wireless broadband remain large cities, especially where high-rises predominate in the downtown business district. In the developed world the largest cities are already fairly well served by fiber for the most part, and fiber, where it is present, is a formidable competitor. The largest business buildings housing the most desirable customers will usually have fiber drops of high-speed fiber rings encircling the city core, and individual subscribers can purchase OC-3 (144Mbps), OC-12 (622Mbps), or, in some cases, wavelength services (variously 1Gbps or 10Gbps). Generally, such customers are lost to wireless service providers because the availability (the percentage of time that a link is available to the user) of the radio airlink will always be less than for fiber, and availability is critically important to most purchasers of high-bandwidth data services.

Also, you cannot discount the generally unfavorable topography represented by most large modern metropolises. Millimeter microwave transmissions demand a clear path to the subscriber terminal, and unless the base station resides on a tower that is considerably higher than any other structure in the vicinity, many promising buildings are apt to remain out of reach within the cell radius swept by the base station. Lower-frequency microwave base stations using non-line-of-sight (NLOS) technology can reach subscribers blocked by a single structure, but there are clear limits in the ability of even the most intelligent adaptive antenna array to lock on a reflected signal that has described several reflections off intervening masonry walls. Whatever part of the spectrum one chooses to inhabit, wireless broadband is hard to employ in large cities with a lot of tall buildings. (Sometimes a wireless link makes sense, however, which is covered in later chapters.)

Wireless broadband has been deployed with greater success in smaller cities and suburbs, both because the markets are less competitive and because the geography is generally more favorable. The first point is fairly obvious; secondary and tertiary markets are far less likely to obtain comprehensive fiber builds or even massive DSL deployments because the potential customer base is relatively small and the cost of installing infrastructure is not commensurately cheaper. I will qualify the second point, however.

Suburban settings with lower population densities and fewer tall buildings tend to be friendlier to wireless deployments than dense urban cores simply because there are fewer obstructions and also because a single base station will often suffice for the whole market's footprint. Nevertheless, such environments still present challenges, particularly when millimeter microwave equipment is used. Indeed, I know of no instance where millimeter wave equipment has been successfully deployed to serve a residential market in a suburban setting.

Lower-microwave equipment is much better suited to low-density urban and suburban settings, and thus it will receive more attention in the chapters that follow; however, where equipment is restricted to line-of-sight connections, a substantial percentage of potential subscribers will remain inaccessible in a macrocellular (large-cell) network architecture—as many as 40 percent by some estimates. Advanced NLOS equipment will allow almost any given customer to be reached, but, depending on the spectrum utilized by the network operator and the area served by a base station, coverage may still be inadequate because of range and capacity limitations rather than obstructions. Unquestionably, the new NLOS equipment will permit the network operator to exploit the available spectrum far more effectively than has been possible with first-generation equipment with its more or less stringent line-of-sight limitation. But as the operator strives to enlist ever-greater numbers of subscribers, the other, harder limitations of distance and sheer user density will manifest themselves. Both range and the reuse of limited spectrum can be greatly enhanced by using adaptive-array *smart antennas* (covered in Chapter 4), but such technology comes at a cost premium. Figure 1-1 shows a typical example of an urban deployment.

Rural areas with low population densities have proven most susceptible to successful wireless broadband deployments both by virtue of the generally open terrain and, perhaps more significantly, the relative absence of wireline competition. But because of the extreme distances that often must be traversed, rural settings can present their own kind of challenges and can require the network operator to invest in multiple, long-range "wireless bridge" transceivers, each with its own high-gain antenna.

Whatever the site chosen for the wireless deployment, mapping the potential universe of users, designing the deployment around them, and considering the local topography are crucially important to wireless service providers in a way that they are not to service providers opting for DSL, hybrid fiber coax, or even fiber. However, in the case of fiber, right-of-way issues considerably complicate installation. In general, a wireless operator must know who and where their customers are before they plan the network and certainly before they make any investment in the network beyond research. Failure to observe this rule will almost certainly result in the inappropriate allocation of valuable resources and will likely constrain service levels to the point where the network is noncompetitive.

Figure 1-1. *Deploying wireless broadband in urban areas*

Examining the Maturity of the Technology

IEEE 802.16 is a new standard, and the equipment embodying it is new. Potential users have the right to question the reliability and robustness of infrastructure gear that has yet to prove itself over the course of several years in actual commercial deployments.

Obviously, such hard proof must remain elusive in the near term, but at the same time the purchaser should not conclude that 802.16 is a pig in a poke. IEEE standards work is absolutely exemplary and incorporates the conclusions drawn from exhaustive laboratory investigations as well as extensive deliberations on the part of the leading industry experts for a given technology. IEEE standards are nearly always the prelude to the emergence of mass-produced chipsets based on the standards, and the major chip manufacturers themselves are looking at tremendous investments in development and tooling costs associated with a new standard investments that must be recouped in a successful product introduction.

The effect of shoddy standards work would be to jeopardize the very existence of leading semiconductor vendors, and to date the IEEE has shown the utmost diligence in ensuring that standards are thorough and well founded. Furthermore, the IEEE will not normally issue standards on technologies that are deemed not to have major market potential.

For example, the IEEE has not set a standard for powerline carrier access equipment or free-air optical simply because those access technologies have not demonstrated much immediate market potential.

In short, the creation of an IEEE standard is a serious undertaking, and I have yet to encounter a standard that is essentially unsound. I think 802.16-based equipment will perform as advertised and will give the network operator the tools to launch service offerings that are highly competitive on the basis of sheer performance.

This is not to say that a network utilizing standard-based wireless broadband equipment is bound to succeed. Many other factors figure into the success or failure of a service network, including

- Qualifying the customer base
- Securing adequate funding
- Making appropriate decisions on the service mix
- Marketing and pricing the services effectively
- Managing the operation for profitability
- Attracting and retaining a technical, managerial, and sales staff that can realize the objectives of the leadership

Notwithstanding the still-considerable challenges in building a broadband wireless network, today the proponent of wireless access has a fighting chance in the marketplace, which was not the case in the past. The following chapters indicate how to seize that opportunity.

CHAPTER 2

Architecting the Network to Fit the Business Model

Broadband wireless provides one of many physical-layer options for the operator of a public service network. Furthermore, the different wireless networking technologies themselves exhibit widely varying capabilities for fulfilling the needs and expectations of various customers and enterprises. More than previous wireless standards, 802.16 addresses a multitude of needs for the users of broadband access services, but it is not invariably the best solution for delivering broadband services in every market.

Broadband Fixed Wireless: The Competitive Context

This section strives to answer the question, when is 802.16-based equipment appropriate? It is the first and most crucial question network operators have to ask themselves when considering the broadband wireless option.

At the risk of stating the obvious, I will enumerate the rival competitive access technologies for broadband before discussing their competitive positioning vis-à-vis wireless.

In the metropolitan space, wireless broadband competes with the following:

- T1 and E1 data services over legacy copper where aggregations of ordinary twisted pairs form the physical medium of propagation.
- Data services based on Synchronous Optical Network (SONET) and Synchronous Digital Hierarchy (SDH), running over fiber linkages.
- Frame relay services running over fiber or T1/E1.
- Ethernet data services running over active fiber-optic linkages.
- Ethernet data services running over passive optical networks (PONs).
- IP data services over active fiber.
- Asynchronous Transfer Mode (ATM) services over active fiber.
- ATM over passive fiber.

- Wavelength services over active fiber.
- Ethernet services over hybrid fiber coax.
- Digital subscriber line (DSL); most existing DSL networks contain components based on ATM and Internet Protocol (IP) as well as Ethernet.
- Powerline carriers where AC transmission lines carry packet data.
- Broadband satellite.
- Free-air or free-space optics where laser beams transmit information over an airlink, dispensing with fiber.
- 2.5-generation (2.5G) and 3G mobile data services, including HSDPA.
- Integrated Services Digital Network (ISDN), a nearly obsolete type of medium-speed data service utilizing double pairs of ordinary copper phone lines to transmit data.
- Storage-area networks represent a special case; most use specialized data protocols of which Fibre Channel, which runs over optical fiber, is the most popular.

Of these rivals, several are currently entirely inconsequential. Broadband as opposed to medium-speed satellite services scarcely exists as yet, and powerline carrier services and PONs are scarce as well, though both appear to be gathering impetus. Pure IP and Ethernet metro services over fiber are growing in acceptance, but they are not well established, and ISDN has almost disappeared in the United States, though it lingers abroad. Finally, free-space optics have achieved very little market penetration and do not appear to be poised for rapid growth. Other services mentioned previously—such as wavelength, 3G mobile, direct ATM services over active fiber, and metro Ethernet over active fiber—have some presence in the market but are spottily available and limited in their penetration thus far.

In this context, broadband wireless does not look nearly as bad as detractors would have it. If you consider the whole array of competing access technologies, broadband wireless has achieved more success than most. Still, it faces formidable competitors among the more established technologies, and these are T1/E1 (including fractional and multiple T1/E1), frame relay, DSL, and cable data.

Among the incumbent technologies, cable data and DSL are the leading technologies for residential services, and business-class DSL, T1/E1, and frame relay are the dominant service offerings for small- and medium-sized businesses. The largest enterprises that require large data transfers tend to prefer higher-speed optical services using both packet and circuit protocols.

Circuit-Based Access Technologies

Within the enterprise data service market, T1, fractional T1 (E1 elsewhere in the world), and business-class DSL are the most utilized service offerings, along with frame relay, which is chiefly used to link remote offices and occupies a special niche.

T1 is usually delivered over copper pairs and is characterized by high reliability and availability, reasonable throughputs, 1.5 megabits per second (Mbps), and inherent quality of service. Its limitations are equally significant. T1s cannot burst to higher speeds to meet momentary needs for higher throughputs, and they are difficult to aggregate if the user wants consistently higher throughput speed. T1s are also difficult and expensive to provision, and provisioning times are commonly measured in weeks. Finally, T1 speeds are a poor match for 10 base T Ethernet, and attempts to extend an enterprise Ethernet over a T1 link will noticeably degrade network performance.

Because it is circuit based and reserves bandwidth for each session, T1 offers extremely consistent performance regardless of network loading. Maximum throughput speeds are maintained at all times, and latency, jitter, and error rates are well controlled. Were the bandwidth greater, T1s would be ideal for high-fidelity multimedia, but, as is, 1.5Mbps is marginal in that regard.

T1/E1 is legacy access technology. The basic standards were developed in the 1960s, and the SONET and SDH optical equipment supporting massive T1/E1 deployments dates back 20 years. In terms of performance level, T1/E1 is essentially fixed, a fact that will put it at an increasing disadvantage to newer technologies, including broadband wireless. Also, the infrastructure for these circuit-based access networks is expensive to build, but, since most of it has already been constructed, it is by now fully amortized.

I do not expect a lot of new copper to be built except in developing countries, and so the last-mile access for T1/E1 must be considered a fixed asset at this time. But, somewhat surprisingly, the sales of SONET and SDH equipment for the metro core have been increasing rapidly through the late 1990s and the opening years of this century, and they are not expected to peak until 2007. Therefore, SONET and the T1 service offerings it supports will be around for a long time.

Prices in the past for T1s were more than $1,000 per month, but they have dropped somewhat, and they are now about $300 to $400 in the United States, though prices vary by region and by individual metropolitan market. Compared to newer access technologies, T1 does not appear to represent a bargain, but it is all that is available in many locales. Moreover, the incumbent carriers that provision most T1 connections are in no hurry to see it supplanted because it has become an extremely lucrative cash cow.

Because of the apparently disadvantageous pricing, T1 services may appear to be vulnerable to competition, but thus far they have held their own in the marketplace. Ethernet and IP services, whether wireless or wireline, will probably supplant circuit-based T1 in time, but as long as the incumbent telcos enjoy a near monopoly in the local marketplace and are prepared to ward off competition by extremely aggressive pricing and denial of central office facilities to competitors, the T1 business will survive. I suspect that T1 connections will still account for a considerable percentage of all business data links at the end of this decade.

Frame Relay

Frame relay is a packet-based protocol developed during the early 1990s for use over fiber-optic networks (see Figure 2-1). Frame relay permits reservation of bandwidth and enables tiered service offerings, but it is not capable of supporting quality-of-service (QoS) guarantees for multimedia, as does ATM, or some of the ancillary protocols associated with IP, such as Multiprotocol Label Switching (MPLS), Reservation Protocol (RSVP), and DiffServ. Also, frame relay does not permit momentary bursting to higher throughput rates or self-provisioning. Frame relay is rarely used to deliver multimedia and other applications demanding stringent traffic shaping, and it is never used to deliver residential service. Usually, frame relay is employed to connect multiple remote locations in an enterprise to its headquarters, and connections over thousands of miles are entirely feasible. Frame relay switches or frame relay access devices (FRADs) are usually accessed from an office terminal via a T1 connection, though other physical media may be employed, including wireless broadband. Frame relay transmissions over long distances, commonly referred to collectively as the *frame relay cloud*, invariably travel over fiber and are usually encapsulated within ATM transmissions.

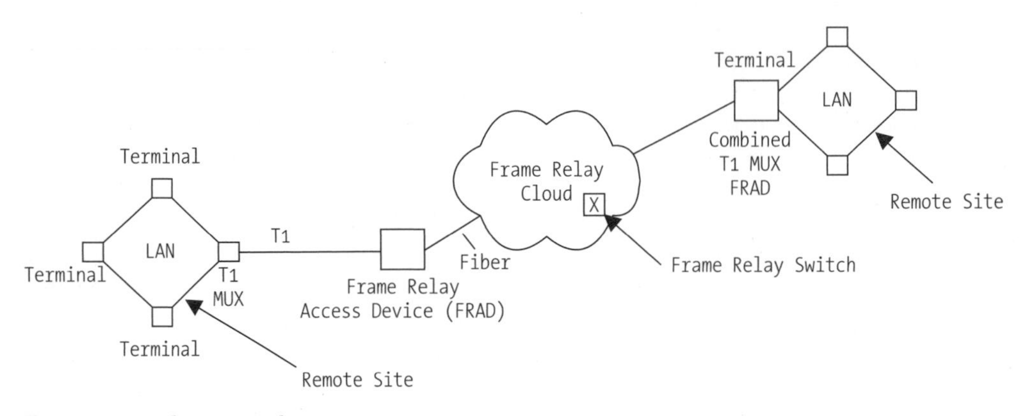

Figure 2-1. *A relay network*

Frame relay services are largely the province of incumbent local phone companies or long-distance service providers. Throughputs vary but are commonly slower than a megabit per second—medium speed rather than high speed. As is the case with T1, frame relay is a legacy technology, standards have not been subject to amendment for years, and not much development work is being done with frame relay devices. The performance of frame relay is not going to improve substantially in all likelihood. Pricing is in the T1 range, with higher prices for higher throughput rates and special value-added services such as Voice-over Frame Relay (VoFR). Also, provisioning of multiple remote locations can be prohibitively expensive with conventional frame relay equipment because the networks do not scale well, and this may limit the popularity of frame relay in the future. Frame relay does not directly compete with wireless broadband in the metro, and thus targeting existing customers for the service makes little sense. Frame relay will continue to lose ground to enhanced metro Ethernet and IP services.

DSL

DSL is arguably the strongest competitor to 802.16 wireless among broadband access technologies. DSL comes in many variants, including asymmetric DSL (ADSL), symmetric DSL (SDSL), G.lite, single-pair high-speed DSL (SHDSL), and very high data rate DSL (VDSL). The distinguishing features of the various substandards are not particularly germane to this discussion and have to do with the speed of the connection and the apportionment of available spectrum upstream and downstream.

DSL utilizes digital signal processing and power amplification to extend the frequency range of ordinary twisted-pair copper lines that were originally designed to carry 56-kilobit voice signals and nothing faster. Aggressive signal processing applied to uncorroded copper can best this nominal limit by orders of magnitude. Commercially available systems can now achieve speeds in excess of 100 kilobits per second over distances of a couple of thousand feet, though the norm for VDSL2, the fastest standards-based DSL technology, is less than 30Mbps over distances exceeding 5,000 feet.

DSL, unlike frame relay, is strictly a physical-layer technology and can be used in tandem with various higher-layer protocols, including circuit digital, Ethernet, ATM, frame relay, IP, and MPEG video, though ATM, IP, and Ethernet are most common today. DSL can support high-quality multimedia if throughput is sufficient and error rates well controlled, but the consistent achievement of high throughput rates is difficult if not impossible in many copper plants.

Plainly put, a DSL network overlay is highly dependent on the condition of the existing copper telephone lines because no one is going to assume the expense of rewiring a phone system—a move to pure fiber would make more sense if that were required. In the presence of corroded copper, both the speed and distance of DSL transmissions are diminished (the best-case distance for moderate-speed transmissions is a little more than 20,000 feet). If the copper plant is compromised, the network operator has no choice but to shorten the distance from the subscribers to the nearest aggregation points known as *digital loop carriers (DLCs).* And since the latter are expensive to site and construct and require costly fiber-optic backhaul to a central office, they can burden the operator with inordinately high infrastructure costs if they are numerous. Nevertheless, SBC has announced an aggressive VDSL2 build-out.

Assessing the cost competitiveness of DSL vis-à-vis other broadband access technologies is difficult because it is so dependent on contingencies. The pricing structure for a carrier owning the copper lines and central office facilities is entirely different from that of a DSL startup obliged to lease copper as well as equipment space in a telco central office. A DSL network is certainly less expensive than new fiber construction because it leverages existing infrastructure, but it still requires a great deal of new equipment and frequently necessitates installation visits to the customer premises by field technicians.

Despite these limitations, DSL services have been expanding rapidly all over the developed world, with especially extensive deployments in East Asia and the United States. In the United States, DSL has found large and growing markets among small businesses and residential users.

To a limited extent, DSL has been used to deliver video services to homes, but the primary offering is simple Internet access. In neither the residence nor the small enterprise are value-added services yet the norm.

Typical speeds for residential service are in the low hundreds of kilobits and slightly higher in the case of business-class services. Some business-class services also offer service agreements in regard to long-distance transmissions over the Internet.

VDSL and VDSL2, the high-speed variants, have the speed to enable advanced IP and Ethernet business services and high-quality converged residential services and, to that extent, must be regarded as a technology to watch. The distances over which VDSL can operate are relatively short, however, little more than a mile best case, and VDSL networks require extensive builds of deep fiber. Only a fairly small number of such networks exist in the world today, though the technology is finding acceptance in Europe. New low-priced VDSL modems are coming on the market that could speed the acceptance of the service somewhat, but that will not reduce the cost of the deep fiber builds necessary to support it.

DSL is a new rather than a legacy technology, emerging from the laboratory about a decade ago (though not subject to mass deployments until the turn of the century), but already DSL appears to be nearing the limits of its performance potential. Where wireless and optical transmission equipment have achieved orders of magnitude gains in throughput speed over the past ten years, DSL has not improved much on the speeds reported years ago. DSL may not be positioned to compete effectively in the future against other access technologies that have the potential for significant further development.

I think DSL is a transitional technology, one that was developed primarily to allow incumbent telcos to compete in the high-speed access market without having to build completely new infrastructure. I further think broadband wireless, as it continues to improve, will become increasingly competitive with DSL.

Finally, basic DSL technology was developed by Belcore, the research organization serving the regional Bell Operating Companies (RBOCs), and was initially intended to support video services over phone lines, services that would enable the RBOCs to compete with the cable television companies in their core markets. The first serious rollouts of DSL were initiated by independents, however, chief among them Covad, Rhythms, and Northpoint, all of which went bankrupt. Independents owned the actual DSL network elements but were obliged to lease lines and locate DSL aggregators (DSLAMs), switches, and routers in central offices belonging to incumbent telcos, generally RBOCs. Such collocation placed the independents in what was in effect enemy territory and left them vulnerable to delaying tactics and even outright sabotage. Dozens of successful legal actions were launched against RBOCs on just such grounds, but the RBOCs simply paid the fines and watched the independents expire.

The wireless broadband operator should draw two lessons from this. First, do not enter into service agreements with competitors, if possible. Own your own infrastructure, and operate as a true independent. Second, realize that the incumbent telcos are grimly determined to defend their monopoly and will stop at nothing to put you out of business. In the past, wireless has not posed a sufficient threat to RBOCs to arouse their full combativeness, but that will change in the future.

Hybrid Fiber Coax

The final major competitive access technology extant today is hybrid fiber coax, the physical layer utilized by the multichannel systems operators (MSOs), industry jargon for the cable television companies (see Figure 2-2). Hybrid fiber coax consists of a metro core of optical fiber that frequently employs the same SONET equipment favored by the RBOCs along with last-mile runs of coaxial television cable. Each run of cable serves a considerable number of customers—as few as 50 and as many as several thousand. The coaxial cable itself has potential bandwidth of 3 gigahertz, of which less than a gigahertz is used for television programming. Most cable operators allocate less than 20MHz of bandwidth to data. Industry research organization Cable Labs is currently at work on a new standard that is intended to exploit the full potential of coaxial copper and to achieve at least an order of magnitude improvement in data speed. Should low-cost, standards-based equipment appear on the market supporting vastly higher throughputs, then the competitive position of cable will be considerably enhanced. In the past cable operators have proved more than willing to make large investments in their plants to launch new types of services. Wireless broadband operators as well as others embracing competitive access technologies would be well advised to watch their backs in respect to cable. Cable is unlikely to stand still in the midterm.

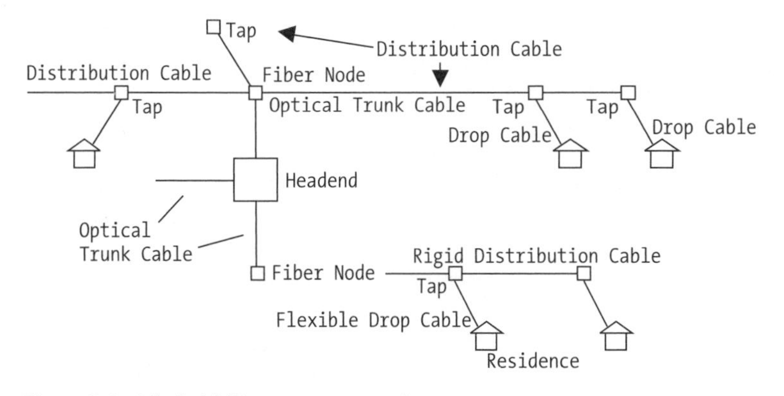

Figure 2-2. *A hybrid fiber coax network*

The speed of a single coax cable far exceeds that of a DSL-enhanced copper pair, but since its capacity is divided among a multitude of subscribers, the speed advantage is manifested to the end user only when the subscriber population on each cable run is small. Cable companies of late have been tending to restrict the number of customers per coaxial cable, but such a strategy is costly and will be pursued only with reluctance and with clear profit-making opportunities in view.

Cable data services aimed at the mass market date back to 1997 in the United States and today account for most of the residential broadband in this country, with DSL ranking a distant second. Unlike the case with DSL, cable data access services are nearly always bundled with video and, increasingly, with cable-based telephone services and video on demand. Cable offers by far the richest service packages for the residential user, and historically the industry has demonstrated a strong commitment to expanding the number of services to cement customer loyalty.

Cable services have historically garnered low customer satisfaction ratings, however, and in truth the actual networks have been characterized by low availability and reliability and poor signal quality. These attributes, it should be noted, are not the consequence of deficiencies in the basic technology but are simply because of the unwillingness of many cable operators to pay for a first-rate plant. Broadband access competitors should not be deceived into thinking that cable systems are consistent underperformers.

MSOs have made some efforts to court business users but have been less successful than DSL providers in signing small businesses. Cable does not pass the majority of business districts, and the cable operators themselves are often not well attuned to the wants and needs of the business customer. Nevertheless, some MSOs have pursued business customers aggressively, and the industry as a whole may place increasing emphasis on this market to the probable detriment of other broadband access technologies. Already several manufacturers have developed platforms for adapting cable networks to serve business users more effectively; these include Jedai, Narad, Advent, Chinook, and Xtend, among others. Cable operators themselves are also beginning to buy the new generation of multiservice switching platforms for the network core that will enable them to offer advanced services based on the Ethernet and IP protocols.

Finally, I should mention *overbuilders*, the competitive local exchange carriers (CLECs) of the cable world. These companies are committed to building the most up-to-date hybrid fiber coax networks, and most of them are actively pursuing business accounts. The ultimate success of the overbuilders is unknowable—they are competing against strongly entrenched incumbents in a period where the capital markets are disinclined to support further ambitious network builds—but in marked distinction to the case with the telco CLECs, the major overbuilders have all managed to survive and to date are holding their own against the cable giants.

Placing cable data access within the context of other competing access technologies is somewhat difficult. Cable could be said to represent the repurposing of legacy technology just as is the case with DSL, but the basic cable plant has been transforming itself so rapidly over the past 15 years that such an interpretation seems less than fully accurate. It is more accurate to say that the hybrid fiber cable plant is an evolving technology that is arguably on the way to having an all-fiber infrastructure in the far future. It may be that local telephone networks will trace a similar evolutionary course—the incumbent local exchanges (ILECs) have been issuing requests for proposals for passive optical networking equipment that would deliver fiber to the curb or fiber to the home—but cable networks lend themselves much more readily to a full conversion to fiber than do telephone networks. My guess is that cable operators will convert more quickly than their telco counterparts. To the extent that this is true, cable emerges as by far the most formidable competitive access technology, and it is one that is likely to preempt broadband wireless in a number of markets. Cable offers potentially superior speed over DSL (and certainly to legacy circuit services); near ubiquity; fairly cost-effective infrastructure; a range of attractive service offerings; and a wealth of experience in network management.

Predicting the course that technological progress will take is difficult, but in the long term, extending into the third and fourth decades of this century, pervasive fiber will establish itself throughout the developed world, packet protocols will be ubiquitous at all levels of the network, and the resulting converged services or full services network will essentially be an evolved cable network. The underlying distinctions between cable and telecommunications networks will vanish, and there will be only one wireline technology.

Wireless Broadband

So where does all this leave wireless broadband?

The singular strength of wireless broadband access technologies is the degree to which they lend themselves to pervasive deployments. At least in the lower frequency ranges, building a wireless network is largely a matter of setting up access points. Subscriber radio modems are destined to decline in price over the course of this decade and will eventually take the form of specialized software present as a standard feature in nearly all mass-market computing platforms, a trend that further supports pervasiveness. Wireless networks will increasingly be characterized by their impermanence and flexibility, and as such they will be complementary to expensive albeit extremely fast fiber-optic linkages.

This book, however, focuses on the present, and current wireless broadband networks are not and cannot be completely pervasive, and subscriber modems are not extremely inexpensive, though they are falling in price. In terms of price and capabilities, wireless broadband is competitive with T1, DSL, and cable; it is better in some respects and inferior and others, and it is highly dependent on accidents of topography as well as local market conditions.

This section first describes how wireless speed and capacity compare with those of the major wireline competitors.

Wireless networks are more akin to cable than to DSL or circuit because they are essentially *shared resource* networks. The same spectrum at least potentially serves every customer within reach of a base station, and the network operator depends entirely on the media access control (MAC) layer of the network to determine how that spectrum is allocated. A network operator could choose to make the entire spectrum available to one customer, but in most cases no single customer is willing to pay a sum sufficient to justify the exclusion of every other party. When a network operator does find a customer who wants to occupy the full spectrum and is willing to pay to do so, the operator will usually link the customer premises with the base station via two highly directional antennas so that the spectrum can be reused in a nearby sector.

How much spectrum the operator has available determines the network's capacity. In the lower microwave region, 100 megahertz (MHz) constitutes a fairly generous allocation, and 30MHz is about the minimum amount necessary to launch any kind of broadband service. You can derive ultimate data throughput rates by utilizing the correct bits-to-hertz ratio. In the lower microwave region, current generation equipment can manage 5 bits per hertz under strong signal conditions, and that number may be expected to rise over time. Thus, the total bandwidth available to most 802.16a operators will allow a throughput of, at most, a few hundred megabits per second—much more than is the case for an individual DSL line or a T1 circuit, but substantially less than can be delivered over hybrid fiber coax. Bear in mind, however, that in a VDSL2 installation, each DSL line can deliver up to 100Mbps, and a network may have thousands of separate lines, so the comparison is not entirely apt. In general, 802.16a is disadvantaged in respect to total throughput as compared to any of the technologies with which it is competing, with the exception of 3G mobile wireless.

Usually, 802.16 radios (which operate in the higher microwave regions) can take advantage of more generous spectral allocations—several hundred megahertz to more than a gigahertz. And where the full bandwidth is used, which it often is, 802.16 radios can vie with fiber. You should understand, however, that the bits-per-hertz ratio is generally poorer for higher-band microwave equipment—often no more than 1 bit per hertz, so the generous bandwidth allocations are to some extent offset by the limitations of the equipment in terms of spectral efficiency. Rather recently, spectrum has been opened above 50GHz in the United States where spectral allocations in the gigahertz are available, and these offer the possibility of truly fiberlike speeds. Then, too, it is not difficult to use such microwave equipment in tandem with free-air optical equipment that is also capable of throughputs of gigabits per second. Thus, aggregate throughputs of 10 gigabits per second (Gbps) could be achieved through an airlink, the equivalent of the OC-192 standard for single wavelength transmissions over fiber.

You must balance these manifest advantages against the poorer availability of the airlink compared to that of fiber. Not a tremendous number of enterprise customers are interested in very high speed and only moderate availability. Usually, customers for high-throughput connections want a very predictable link.

The overall capacity of a broadband wireless network as opposed to the maximum throughput over any individual airlink is largely a function of the number of base stations in the network. The denser the distribution of base stations, the more customers can be accommodated—provided, that is, that the power of individual transmissions is strictly controlled so that transmissions within each individual cell defined by the base station do not spill over into adjacent cells. The situation is almost analogous to cable networks where capacity can be increased by almost any degree simply by building more subheadends and more individual runs of coaxial copper.

The problem, of course, is that base station equipment is expensive, and leased locations on rooftops or the sides of buildings can represent heavy recurrent costs. And unless a base station is equipped with an adaptive array antenna, it cannot remotely compare in capacity with a cable subheadend that can easily accommodate many thousands of users. You should remember that the first use of broadband wireless in the United States was to distribute "wireless cable" television programming over approximately 200MHz of spectrum, a fairly generous allotment. Even so, wireless cable service providers could not compete effectively with conventional cable operators, and most went out of business within a few years.

I should point out that the difficulty in reaching all potential customers is much greater with higher microwave operators covered by 802.16. Any building that is not visible from the base station is not a candidate for service—it is that simple—so the capacity of any individual base station is quite limited. Unfortunately, base station equipment intended to operate at these frequencies is usually far more costly than gear designed for the lower microwave region, so the construction of additional base stations is not undertaken lightly.

Another extremely important limitation on capacity affecting upper microwave operators is the nature of the connection to the individual subscriber. Upper microwave services are invariably aimed at business users. Because of line-of-sight limitations, the operator looks for a lot of potential subscribers within a circumscribed space—a business high-rise is the preferred target. Instead of providing a subscriber terminal to every subscriber—which is a prohibitively expensive proposition because subscriber terminals for these frequencies are nearly as expensive as base station equipment—the operator strives to put a single terminal on the roof and then connect customers scattered through the building via an internal hardwired Ethernet, though a wireless LAN could conceivably be used as well.

Distributing traffic to subscribers may mean putting in an Ethernet switch in an equipment closet or data center, and it will undoubtedly involve negotiations with the building owner, including possibly even recabling part of the building. Successfully concluding negotiations with the owners of all the desirable and accessible buildings is a difficult undertaking, and the network operator is unlikely to be wholly successful there. In contrast, the incumbent telco offering high-speed data services will already have free access to the building—the real estate owner can scarcely deny phone service to tenants and expect to survive—and thus has no need to negotiate or pay anything. Owners can simply offer any services they see fit to any tenant who wants to buy it.

Such real estate issues do not constitute an absolute physical limit on capacity, but in practical terms they do limit the footprint of the wireless network operator in a given metropolitan market. One has only to review the many failures of upper microwave licensees to build successful networks in the United States to realize that despite the ease of installing base stations compared to laying cables, microwave operators do not enjoy a real competitive superiority.

As was the case with throughput, so it is with total network capacity. Whatever the frequency, wireless broadband does not appear to enjoy any clear advantage.

I have already discussed availability and reliability. Wireless will always suffer in comparison to wireline, and wireless networks are apt to experience a multitude of temporary interruptions of service and overall higher bit error rates. In their favor, wireless networks can cope much better with catastrophic events. Though an airlink can be blocked, it cannot be cut, and the network can redirect signals along circuitous paths to reach the subscribers from different directions in the event of strong interference or the destruction of an individual base station. This is not necessarily possible with most equipment available now, but it is within the capabilities of advanced radio frequency (RF) technology today.

In terms of QoS and value-added services, broadband wireless is getting there, but it is generally not on par with the wireline access media. The first generation of 802.16a equipment cannot emulate the cable networks in simultaneously offering video programming, voice telephony, and special business services.

So what is the competitive position of wireless broadband, and where does it have a chance of success?

Many individuals with experience in wireless broadband have concluded that wireless networks should be contemplated only where the leading wireline access technologies such as T1/E1, frame relay, DSL, and cable data are not already established. Each of these access methods is fairly well proven, the argument goes. Moreover, the plant is paid for; that is, the infrastructure is fully amortized. Incumbents offering these services can and will temporarily slash prices to quash competitors, so the wireless operator cannot necessarily compete in price even in such cases where wireless infrastructure can be shown to be much more cost effective.

Still, I am not so pessimistic as to concur with the position that wireless cannot compete against wireline under any circumstances and that it must therefore enter only those markets where there is nothing else. The case for wireless is not nearly so hopeless. A great multitude of business customers, perhaps the majority, is badly underserved by the T1 and business DSL services that prevail in the marketplace today. They are both slow and relatively expensive, and the service providers make little effort to make available value adds such as storage, security provisions, application hosting, virtual private networks (VPNs), conferencing, transparent LAN extensions, self-provisioning, and bandwidth on demand. T1 services, to put the matter bluntly, are a poor value. None of these limitations apply to the wireless broadband networks of today, and the wireless operator can compete against incumbent wireline services simply by offering better and more varied service offerings and by being more responsive to subscriber demands—this despite the fact that wireless equipment lacks the capacity or the availability of wireline networks.

Competing for residential customers is much more difficult. Wireless broadband cannot transmit video as effectively as can cable and cannot ensure the same level of availability for voice as twisted-pair copper, so it lacks a core service offering. All it can provide in the way of competitive services is Internet access and value-added business services. Wireless can, however, afford the user a measure of portability and even mobility, and that may actually be its principal selling point in respect to residential customers. Metricom, the first company to offer public access over unlicensed frequencies, achieved some measure of success by emphasizing portability and later full mobility.

Determining When Broadband Wireless Is Cost Effective

Network operators contemplating a wireless access approach in a given market should consider not only the initial purchase price of the infrastructure but the following factors:

- Total cost of ownership for the wireless components of the network
- Scalability of the networks, both in regard to the absolute number of subscribers who can be served and the marginal costs associated with adding customers

- Subscriber density; the proportion of potential customers who are physically reachable
- Ease and cost of deployment; the speed and availability of the connections
- Types of services that can be supported
- Speed of service provisioning
- Degree to which the network operator will depend on other service providers

Total Cost of Ownership

When the first broadband wireless networks appeared in the late 1990s, proponents liked to emphasize the cost advantage of a wireless solution. If one did not have to lay cable, which was clearly the case with wireless, then one neatly avoided the enormous installation costs associated with digging up the street and encasing the cable in conduit or, in some cases, in concrete channels, as well as the expense of running cable through buildings to the final destination at an Ethernet or ATM switch. Wireless broadband appeared to enjoy a clear and indisputable advantage, and who could possibly suggest otherwise? Several years and hundreds of business failures later I can say only that wireless is less expensive and more cost effective in some individual markets and not in others.

Wireless indeed eliminates cabling in the final connection between the operator-owned network node and the subscriber terminal and in that manner avoids a highly significant cost factor. Cable excavation and installation can run anywhere from a few thousand dollars a mile in rural areas to more than a million dollars in large cities, with median costs running in the tens of thousands of dollars per mile (these figures are based on my own primary research). But such comparisons ignore that two of 802.16's most formidable competitors, DSL and cable data, commonly use existing cable infrastructure and do not require cable installation. Such is frequently the case with fiber-optic services as well, with fiber operators striving to secure unused "dry fiber" from public utilities or, alternately, running new fiber through existing passageways such as sewers or gas pipes. In most cases, fiber access networks require new builds and are inordinately expensive, but not invariably, and any wireless operator who assumes a competitive advantage over a fiber-based service provider because of cost without any clear evidence to support that assumption may be sadly mistaken.

Another major cost factor that was initially ignored in most cost estimates for broadband installations is the so-called truck roll, industry slang for the professional installation of the subscriber terminal. Except in cases where a connection to a subscriber terminal is already in place waiting to be activated, or where a portable wireless device such as a cell phone is involved, something has to be done to make the connection to the network. Generally in the case of broadband wireless, that something will involve the installation of at least two devices on the subscriber premises: an antenna to transmit and receive signals and a modem on the subscriber's computer, either external or card based, along with appropriate software.

Because of the vagaries of subscriber terminals, the unfamiliarity of many users with broadband services, and the lower degree of automation in earlier broadband access equipment platforms, most broadband service providers in the past elected to send technicians out to perform the installation and make sure that the service was adequate. Such is still almost invariably the case with very high-speed services offered to large enterprise users.

As broadband operators began to examine their operational expenses, they realized that truck rolls were critically expensive, often running as much as $1,000 per subscriber when all associated expenses were taken into account. Because of the normal problems arising when almost any new, complex, and unfamiliar technology is introduced into a computing environment, many customers required two or more truck rolls, each costing approximately as much as the first.

Obviously this was a serious problem, one that detracted greatly from the profitability of early broadband services. If, for example, a customer in a residential installation was charged $50 per month and had $3,000 of expenses in the installation process, years would be required to recoup expenses, and that was assuming the customer was not induced to try a competing service.

Each kind of broadband service poses its own peculiar installation problems, as it happens. Fiber demands precise trimming and alignment of the optical fibers themselves, DSL requires testing and qualifying every copper pair, and cable requires various nostrums for mitigating electrical noise and interference. With wireless it is primarily positioning and installing the antenna. Wireless may generally be said to pose the greatest installation difficulties, however, at least in terms of the subscriber terminal.

In some cases involving a wireless installation, the installation crew has to survey a number of locations on or in a building before finding one where the signal is sufficiently constant to permit operation of a broadband receiver, and that process can consume an entire workday. Rooftop installations serving several subscribers, which are commonplace in millimeter microwave installations, require the construction of large mountings and extended cable runs back to subscriber terminals and therefore can be very expensive. And, worst of all, the RF environment is dynamic, and an installation that experienced little interference at one time may experience a great deal at some future time—necessitating, you guessed it, another truck roll.

With all current millimeter microwave equipment and first-generation low microwave components, one could expect difficult and costly installations a good deal of the time. Second-generation non-line-of-sight lower microwave equipment, on the other hand, is usually designed to facilitate self-installation and indoor use, eliminating the truck roll in many cases and appearing to confer a decisive advantage on wireless. But unhappily for the wireless operator, cable and DSL have their own second generations, and self-installation of either technology is fast becoming the norm. Furthermore, self-installation normally confers no performance penalty in the case of cable or DSL, but with wireless, this is not the case. With an indoor installation (the easiest type to perform because the user is not obliged to affix mountings on outside walls or roofs), the effective maximum range of the link is much reduced, and the network operator is consequently obliged to build a denser infrastructure of base stations. In some instances this considerable added expense may be offset by the reduction in truck rolls, but not always.

Incidentally, regarding this matter of truck rolls, outsourcing is rarely advisable. Companies that maintain their own crews of technicians can sustain installation costs at a fraction of those charged by contractors—something to think about when planning a rollout. Assembling and training installation crews may be time consuming, but outsourcing may simply be cost prohibitive.

Setting up wireless base stations must be viewed as another major cost factor in building a wireless broadband network. Of course, rival technologies must also set up large network nodes that are the equivalent of broadband wireless base stations, but in the case of cable and DSL these usually already exist, the headend and the central office, respectively. In other words, they do not have to be built from scratch. Furthermore, the capacity of major DSL and cable aggregation nodes is generally much greater than is the case for wireless base stations. A central office in a DSL network can handle thousands of lines, as can a cable subheadend, that is, the termination for the coaxial copper cable connecting the cable customers. Depending on the amount of spectrum available to the wireless operator, an individual base station is more likely to serve hundreds or dozens of subscribers rather than thousands.

Wireless equipment for base stations has come down considerably in price to the point where it is quite cost competitive with cable and DSL equipment, particularly in the lower microwave regions, but equipment may be only a fraction of the total cost of a base station. In very few instances does the network operator own the various locations where the base stations are to be sited. Thus, roof rights and other right-of-way arrangements have to be negotiated with real estate owners, and these are generally recurrent costs. It is difficult to generalize about the cost of such leases, and it is best to map the network and secure all necessary rights of way before going further. If the cost of doing so appears likely to be exorbitant, a wireless network simply may not be feasible within that given market.

Although many details should be considered when estimating total cost of ownership, the way such information is used is fairly straightforward. Operators first have to determine the pricing of services that will make them competitive with other broadband service offerings in the area and then decide if the cost of purchasing, leasing, and maintaining the infrastructure can be borne with the revenues from competitively priced services while leaving something for profits.

How Scalable?

Scalability of the network, the second factor, is no less important than total cost of ownership, because it determines the long-term prospects of the network operator.

In terms of the ability of the central network management software to handle multiple base stations and multitudes of customers, the scalability of 802.16-based networks is not a problem. The real issue is how many base station sites can be secured and what arrangements the operator can make for backhaul. Theoretically, a large network should be more profitable to operate than a small one because central office costs are relatively fixed, as are access costs for an Internet point of presence. Often, however, the network operator will be able to identify only a few buildings with good sales potential and may not be able to sign up more than a few hundred customers. The question then becomes, how does the network scale down? Unless the network operator can upsell the customer on a lot of value-added services and applications, the answer to that question may not be reassuring because the fixed costs of running the network must be amortized among a relatively small number of customers.

Regarding the number of potential customers that are reachable, I have already touched upon this topic. With the new non-line-of-sight equipment for the lower microwave region, most potential customers can now be reached—if they are in range of a base station. With higher-frequency transmissions, the ability to reach buildings remains a problem. The only solution is to utilize a mesh architecture and place routers or switches at each subscriber premises, but no company with the exception of Terabeam is currently making millimeter microwave equipment that can operate in a mesh topology.

Service Delivery and Broadband Wireless

I have also mentioned briefly service offerings. The physical layer of any network is essentially a pipe and should be able to support any type of service or application provided that the raw bandwidth to do so is available. Wireless happens to be unique in that the airlink is intermittent in its capacity, its signal-to-interference ratio, and its susceptibility to fades and interruptions. Such intermittency makes it difficult to deliver certain services, particularly rich multimedia and high-speed, real-time interactive applications.

So what services can wireless 802.16 networks deliver?

Along with basic high-speed Internet access, 802.16 can support the following: VPNs, IP and circuit second-line telephony, telemetry, conferencing, bandwidth on demand and self-provisioning, and storage service networks.

VPNs

VPNs are protected communications going back to a corporate Web site and are demanded by most businesses of any size for employees engaged in telecommuting or remote accessing of corporate data. Several means of enabling VPNs exist. IP/MPLS and Ethernet VPNs form the latest generation of VPN service offerings and can both be managed by the service provider.

IP and Circuit Second-Line Telephony

All 802.16 equipment can support IP telephony, and some can support circuit telephony. Because of the excessive bandwidth demands of circuit telephony and the extremely high cost of traditional class 5 circuit switches, I do not recommend wireless broadband operators offering such legacy services. IP telephony is a different matter because of the much lower cost of the equipment and the high degree of bandwidth efficiency associated with the technology, but you should remember that an airlink cannot ensure the same availability as the copper plant and that radios cannot operate without AC power whereas telephones can. On the other hand, an airlink cannot be cut. Broadband wireless networks are not well suited to offering primary telephone services in markets where such services are already provided over copper, though they may have an application in settings where phone service is otherwise unavailable. If, however, a company chooses to offer such "wireless local loop" services, little network capacity will remain for anything else.

Telemetry

Telemetry is essentially machine-to-machine communication and generally takes the form of remote monitoring. Examples include measuring inventory in vending machines and signaling when restocking is needed, as well as monitoring pipelines for leaks. Wireless is uniquely well suited to telemetry, and it is service that many operators neglect to promote.

Conferencing

Conferencing, particularly videoconferencing, is an application finding increasing acceptance in the enterprise. All 802.16 equipment can support IP-based video and audio conferencing.

Bandwidth on Demand and Self-Provisioning

Bandwidth on demand is a temporary change in the amount of bandwidth or throughput allocated to a subscriber in order to meet an immediate need such as large file transfers or videoconferencing. Self-provisioning allows subscribers to change the terms of their service from a secure Web site without the intervention of a sales agent. Both are more a function of network management than the physical link, and both are possible with 802.16 standards–based equipment. Where bandwidth on demand and self-provisioning have been offered in wireline networks—and they have not been offered by many service providers—they have always proved extremely popular, and I think they are highly desirable service offerings for wireless operators.

Storage Service Networks

Storage is a network application where vital information is off-loaded to remote storage facilities and invoked thereafter, as it is needed.

Subscriber Density

Subscriber density relates pretty directly to system capacity and more directly to how frequently spectrum can be reused. Since frequency reuse is an absolutely key concept to operating any metropolitan wireless network, I will devote some space to the topic in this discussion of subscriber density.

Any given radio frequency can be occupied by only one user within a given propagation path. Two or more users attempting to use the same frequency simultaneously will interfere with one another. To prevent interference the network architect must either assign a single channel to each user or assign recurring time slots to individual users within the same band (combinations of the two approaches are possible as well). The limit of the ability of a given slice of spectrum to carry traffic is reached when every frequency is occupied for every wave cycle.

In practical terms, such a limit can never be reached, but it can be approached through such modulation techniques as Code-Division Multiple Access (CDMA) and orthogonal frequency division multiplexing (OFDM), where individual transmissions are distributed across the entire available spectrum in complex interleavings that leave relatively little spectrum unoccupied for any length of time during periods of heavy network traffic.

Once available spectrum is completely filled, the only way the operator can support more traffic is to employ some means of reusing the spectrum within some restricted area. This is accomplished by two methods: transmitting the signal in a narrow beam by means of a directional antenna and transmitting at low power so that the signal fades to insignificance at distances beyond the terminal for which it is intended.

Directional antennas themselves use two techniques for achieving their directional characteristics: focusing the transmission in a parabolic dish reflector and using complex constructive and destructive interference effects from several omnidirectional monopole antennas to shape a beam. The second type is known as a *phased array* and is far more flexible. Directional antennas ordinarily work only with fixed installations where subscriber terminals do not move in relation to the base station, and thus they impose a limit on mobility or even much portability. They also require careful management because they must continually be realigned as subscribers are added to or dropped from the network.

Directional antennas produce one unfortunate side effect; they extend the reach of the transmitter considerably by concentrating energy along a narrow wave front, and they change the attenuation characteristics of the signal. This means that spill going past the intended subscriber terminal can interfere with distant terminals elsewhere in the network. This can be a real problem in mature networks where the footprint is divided into a series of adjacent cells and the intent is to reuse spectrum from cell to cell (frequencies can rarely be reused within adjacent cells, and wireless networks ordinarily require intervening cells separating those using identical frequencies).

Both parabolic reflector antennas and phased array antennas can be aggregated to produce what are known as *sectorized* antennas—groups of directional antennas distributed on the same vertical axis and dividing the cell defined by the base station into sectors of roughly equal area. Such antennas will permit nearly fourfold increases in spectral efficiency within a cell but will increase interference in adjacent cells by the square of the existing quantity. In other words, they are no panacea, but they may provide the right solution for certain distributions of subscribers.

In the last few years, adaptive phased antenna arrays have been developed where computing engines continually evaluate network conditions and shape the directivity patterns of signals emanating from the array so as to mitigate interference while permitting maximum traffic densities. Alone among directional antennas, adaptive phased arrays can support full mobility in the subscriber terminal.

Adaptive phased array antennas, also known as *smart antennas*, offer other benefits as well, which I will cover in Chapter 4. They constitute what is truly a breakthrough technology that can significantly extend the capabilities of the wireless network and significantly increase both capacity and subscriber density. And yet they have been little employed to date because of the substantial price premiums they have commanded. Prices are beginning to come down, and more companies are entering the field, and within a two- or three-year period such devices will most likely become commonplace. But as of this writing, choices are still limited.

The second technique for achieving high spectral efficiency (using a multitude of low-powered base stations defining *microcells*) will allow the network operator to achieve almost any degree of subscriber density, but at a price. Base stations cost money, and leasing space on which to situate base stations costs more money. The trick in succeeding by subdividing a network into smaller and smaller microcells is determining beforehand how many additional subscribers you are likely to attract. Subscriber growth rarely has a linear relationship with infrastructure growth, and the marginal cost of gaining new customers is apt to increase. Wireless operators seeking customers seldom face an initial situation where the number of customers wanting to be admitted to the network exceeds the amount of network capacity to support them. No broadband wireless operator to date has faced insatiable demand, so network operators should proceed with the utmost caution in building excess network capacity.

Absent adaptive antennas arrays and microcell architectures, the network operators need to deploy fixed antennas carefully based on subscriber growth assumptions for the network and calculate the hard limits of how many subscribers can be allocated how much spectrum. While engineering formulas and design software exist for plotting antenna deployment and directivity for maximum utilization of spectrum within an overall cellular architecture, performing such calculations in the face of uncertainties as to the precise distribution of subscribers at various phases of network expansion is at best an estimate. Chapter 4 discusses such issues in further detail. Here I must emphasize that the first task facing the network operator is to plot the probable distribution of subscribers, with breakdowns as to the relative density of high-value business customers versus basic service subscribers in various locales. Only after that exercise has been completed should infrastructure requirements then be projected.

Local Topography and the Type and Distribution of Man-Made Structures

Of further concern to network operators contemplating a build is not just the density and distribution of potential subscribers but the architecture of the locations where they will be receiving service.

Business parks and high-rises are in many respects the most desirable locations in which to offer service, both because of the abundance of high-value customers and because bandwidth can be aggregated and made available via a single narrow-beam airlink directed to a single rooftop antenna. Distribution to individual subscribers within the complex would then take place through an internal network.

Where such locations are identified, the network operator should secure the right to install an external subscriber terminal prior to designing a base station to serve the building or business park and arrive at an equitable arrangement with the real estate owner to use the internal network. If this cannot be accomplished before the base station is erected, network operators are simply wasting their money.

Subsequent to securing these rights, the network operator should canvass the tenants of the building and determine that an adequate number will subscribe to the service to justify the cost of the installation. The worst thing a network operator can do is embrace the philosophy of "build it and they will come." Countless broadband access providers have failed through just such visionary zeal. Broadband services are a market, not a mission from God. Offer them only where they are truly wanted and where subscribers are willing to support the network.

Multitenant units (MTUs) are the residential equivalent of business parks and high-rises and are subject to many of the same considerations. The difference is that far fewer of them have internal high-speed cabling for distribution of broadband services to the tenants. If an internal network is not present, then the network operator will have to construct it, and that can be extremely expensive and time consuming, especially if coaxial cable is employed.

The problem facing lower microwave wireless operators contemplating MTUs is that they can really compete with DSL only in providing basic high-speed access; they cannot very readily compete with cable in offering video programming because they lack the bandwidth to carry scores or hundreds of channels of video. Millimeter microwave transmissions covered by the 802.16 standard could conceivably serve an MTU market for converged services, but the high cost of the equipment would put the network operator at a competitive disadvantage with cable. In general, MTUs do not comprise an especially favorable market for broadband wireless at this time.

Single-family homes or small businesses located along thoroughfares also constitute a difficult market to serve. Sectoral antennas cannot reapportion spectrum efficiently enough to support high reuse within a broad, low-density suburban setting, so the network operator must restrict throughput to individual subscribers, especially if they are numerous in a given area, in order to preserve scarce spectrum. Microcells provide a partial answer, but since residential users and very small businesses tend to confine themselves to basic service packages, the additional revenues may not justify the outlays.

I do not know of any very successful mass deployment of broadband wireless services to a consumer user population in an urban setting. That is not to say it cannot be done, but the economics of the model do not look favorable with current market conditions and current technology.

A growing phenomenon in the wireless data business is the provision of access services in public spaces by means of *hotspots*, very small radius cells serving casual users. Currently, almost all hotspots utilize 802.11 wireless LAN equipment, not the more expensive 802.16. Although the emerging 802.20 standard may eventually have a place in the hotspot market, I do not see 802.16 equipment being utilized to provide direct services to subscribers. Operators can employ 802.16 links to backhaul hotspot traffic, however.

This brings me to a quite distinct and underappreciated market for wireless services that is presently being met with 802.16 equipment and is undoubtedly accounting for more revenue than any other, namely, backhaul services. *Backhaul* refers to a connection from a base station to a central office, and it is used in mobile as well as fixed-point networks, though far more in the mobile arena. An increasing number of backhaul connections are being made over fiber, but a definite market for wireless backhaul exists and will continue to exist.

Since the subscriber for such services is another service provider, the requirements for the backhaul network are much more stringent than is the case for a network serving enterprise customers, and many startups would not be able to meet such requirements. In any case, wireless backhaul probably represents a declining business. Bear in mind that microwave was formerly the dominant technology for handling long-distance telephone traffic and now has all but disappeared. I predict that fiber backhaul will achieve similar absolute dominance in metropolitan areas.

Speed of Deployment

Ease and speed of deployment, the next consideration facing the prospective network operator, would appear to favor a wireless solution. But this is assuming that negotiations with site owners proceed smoothly. If they do not, the deployment process may become interminable. Securing backhaul may also be a problem, particularly if high-speed wireline links are not in place where they are needed or are priced out of reach. In general, I favor wireless backhaul (transmitting voice and data traffic from a cell to a switch) wherever it is feasible because otherwise the network operator ends up competing with the incumbents on their own turf and, worse, depending on them for facilities.

Independence from Incumbents

This brings me to the final point, the degree to which the wireless network operator can be truly independent from incumbents.

Unlike the DSL service provider, the wireless operator has no reason to collocate in the telco central office. If operators want to provide second-line telephone services, they can buy their own switch, preferably an IP softswitch. They do need a relationship with a long-distance service provider, but that entity need not be a competitor.

Making a Final Determination

Most books about setting up wireless networks deal strictly with logistics. But the network operator should spend an equal or greater amount of time concentrating on the business case and whether a network can in fact be constructed in a given geographical market that sells its services profitably. Going wireless is not just a matter of figuring out how to do it. It is equally a matter of determining why or why not to do it.

CHAPTER 3

Strategic Planning of Spectrum and Services

This chapter enumerates the various steps involved in the strategic planning of a wireless broadband network. The assumption at this point is that the network operator has already performed a market and logistical study of the geographical area in which service is to be offered and has determined the following:

- A considerable unmet demand exists for competitively priced broadband services, and the challenge posed by incumbents is not overwhelming.
- The local political climate is such that obtaining right of way, permitting, and spectrum itself will not pose insurmountable obstacles.
- The basic topography of the region and distribution of potential customers is such that a wireless solution is both feasible and desirable.
- A decisive "first-mover" advantage will accrue to the network operator, rendering effective competition from other wireless service providers unlikely.
- The network will enjoy good long-term growth prospects.
- Adequate backhaul facilities can be secured for both the immediate and future needs of the network.
- Adequate peering relationships with long-distance service providers can be negotiated so that quality of service (QoS) can be maintained in local area network (LAN) extensions, Voice-over IP (VoIP), conferencing, and other low latency or low error rate applications.
- The network can truly own its own infrastructure and not be beholden to competitive incumbent carriers.
- The network will be positioned to adopt foreseeable enhancements in wireless technology without having to resort to "forklift upgrades."

- The organization will be in a position to forge relationships with third-party providers of content, applications, and value-added services.
- A roadmap will be in place for modifying the network over time to support such converged services.

Basically, the network operator will have done everything possible to assess the full market potential of the locale in which the network is to be situated and the logistical requirements for launching the network and will leave as little as possible to chance. Only when such assessments are complete should the operator proceed with the actual planning stage of the rollout.

Selecting the Appropriate Spectrum to Meet the Requirements of the Targeted Customers: Propagation Characteristics Across the Radio Spectrum

Issues of spectrum should be dealt with during the market assessment phase of network planning, at least to the extent of finding out what spectrum is available in the designated market. Simply put, if the requisite spectrum is not available, then the project cannot go forward.

As indicated in Chapter 1, the 802.16 standard covers the bands between 2 gigahertz (GHz) and 66GHz, but other bands are available above and below this region that can be used for providing broadband access (I will cover these as well). As desirable as standards-based equipment may be for the network operator, occupying the spectrum that is most useful is equally desirable.

The 802.16 standard itself divides the radio frequency (RF) spectrum into large, slightly overlapping blocks, the first of which extends from 2GHz to 11GHz and the second from 10GHz to 66GHz. In terms of the propagation characteristics of the signal, these divisions are both too coarse and so arbitrary as to be almost meaningless from an engineering perspective. In regard to network planning, more useful divisions would extend from 700 megahertz (MHz) to 3GHz, then from 3GHz to 10GHz, from 10GHz to 40GHz, and from 40GHz to 100GHz. I should point out here that the IEEE's decisions as to what spectrum to include in the standard are based primarily on what bands are actually available for broadband deployments in the United States and secondarily in other developed nations. They are not based on the natural divisions in the spectrum reflecting the changing nature of wave propagation with respect to frequency.

Before you consider such natural divisions, I will give an overview of the useful spectrum—useful, that is, from a communications viewpoint.

Overview of Commercial Radio Spectrum Suitable for Broadband Data Applications

Radio transmissions occur at frequencies as low as 28 cycles per second (low audio frequencies) and as high as a couple of hundred billion cycles per second (gigahertz). Audio frequency radio transmissions have no commercial application and are currently utilized only by navies for communicating with deeply submerged submarines. On the other hand, transmissions exceeding 100GHz are currently used only for imaging (radio photography). Electromagnetic waves occurring above 300GHz are conventionally considered infrared light, but there is no arbitrary frequency where the characteristics of wave propagation change drastically and the radio wave suddenly assumes the properties of radiant light.

Commercial broadcasts commence at hypersonic frequencies in the hundreds of kilohertz (thousands of cycles per second), frequencies that are assigned to AM broadcast stations in the United States. These frequency bands are wholly unsuitable for high-speed data for a number of reasons. Wavelengths span literally hundreds of yards and require immense amounts of power to propagate at detectable signal levels. And because a radio signal can convey only data rates that are a few multiples of the carrier frequency, such low-frequency signals simply cannot transmit data very quickly.

As you proceed up into the megahertz (millions of cycles per second), the bands become increasingly well suited to the transmission of data, but most of these bands have long ago been assigned to what are now well-entrenched commercial and governmental users and therefore are effectively unavailable. Only as you approach the low microwave region from 1GHz to about 10GHz do bands become available that can, on the one hand, support high-speed data traffic and, on the other hand, have not been assigned to users who are so influential that they cannot be made to surrender the spectrum for new uses.

Much of the vast amount of radio spectrum located between 2GHz and approximately 100GHz lends itself to data transmission simply because high frequencies enable high data throughputs. What is not useful are those regions of the spectrum where atmospheric conditions conspire to limit range. In the following sections, you will examine the microwave region much more closely, and I will discuss the characteristics of those bands that have already been allocated for data use in the United States and elsewhere.

Beachfront Property: The Lower Microwave Frequencies

Spectrum available for high-speed data starts in the ultrahigh frequency (UHF) bands beginning at 300MHz and extending to 3GHz. In the United States the lowest frequencies currently available for broadband wireless transmissions reside between the 700MHz and 800MHz spectrum formerly assigned to television. Further spectrum is available in the United States between 902MHz and 928MHz, at 2.3GHz, at 2.4GHz, from 2.5GHz to 2.7GHz, and in several bands from 5GHz to 6GHz. Bands located at 2.4GHz and at 5.8GHz are widely available across the globe. Throughout most of the world, though not in the United States, a band centered at 3.5GHz is also available for public access data networks and is fairly widely used. Early in 2005 the Federal Communications Commission (FCC) approved new unlicensed spectrum for broadband data services located between 3650MHz and 3700MHz.

The spectrum between 3GHz and 30GHz is termed *super high frequency (SHF)* but is not all of a piece in regard to the characteristics of RF transmissions within this frequency range. Transmissions occurring from 3GHz to approximately 10GHz and occupying the lower third of the SHF region really have more in common with UHF in that they are relatively limited in throughput, do not readily conduce to high degrees of frequency reuse, and, perhaps most important, share a vulnerability to what is known as *multipath distortion.*

Multipath distortion is a condition in which the signal interferes with itself because reflections off physical boundaries converge with the direct signal, causing the signal level at the receiver to swell or fade depending on the phase alignments of the converging waveforms at the moment they interact with one another. In addition, multipath results in errors in the bitstream at the receiver inasmuch as the reflected signals are delayed relative to the direct signal, causing bits to appear out of sequence. Multipath varies enormously according to the position of the transmitting antenna in relationship to the ground and to large obstructions, and according to the position of the receiving antenna in relationship to the direct and reflected signals impinging on it. If the receiving antenna is not in an area where significant cancellations or reinforcements are taking place, reception will be fine. But move it a few inches, and the signal may become almost unrecoverable.

In the past, multipath has been an endemic problem for networks operating in the lower microwave region. This is a problem that could be mitigated but never entirely solved by careful installation and by using *diversity antenna systems* that consisted of two or more spaced antennas and *associated smart circuitry* that would choose the optimal signal—that is, the one least afflicted with multipath. Today various new and sophisticated modulation techniques, as well as adaptive antenna technologies, are emerging that confer a considerable degree of immunity from multipath on broadband receivers and render placement fairly noncritical. The greatest reductions in the effects of multipath are to be had with certain types of adaptive antennas; however, such technology remains expensive to implement and is by no means widely present in the marketplace. One modulation technique, known as *frequency hopping*, provides almost absolute immunity to multipath; however, it is not specified in the 802.16 standard and is not particularly spectrally efficient either.

Multipath afflicts transmissions from 300MHz all the way up to about 10MHz, but as we enter the SHF bands (3GHz–30GHz) a new problem manifests itself: increasing susceptibility to blockage from walls. Beyond about 2.5GHz, transmitting through walls becomes increasingly difficult at reasonable power levels and at reasonable distances. Beyond 3GHz, the problem becomes fairly acute, and in-building antenna mounting becomes essentially impractical for receiving outdoor transmissions in the popular 5.8GHz band.

Another problem appearing as low as 2GHz and worsening progressively with frequency thereafter is vulnerability to signal interruption in the presence of foliage. Trees may be regarded as vessels filled with water, and microwaves tend to give up their energy to water when they encounter it directly—this in fact is the principle behind microwave ovens. A customer whose terminal is blocked by trees is therefore unlikely to be able receive a signal of adequate strength. In such cases the network operator has two choices: elevate the transmitting antenna sufficiently that the signal clears the treetops or see whether the owner of the trees can be prevailed upon to trim or remove them.

In spite of the vulnerability of lower microwave transmissions to physical obstructions, spectrum in this region, especially below 4GHz, is extremely valuable, affording the user a combination of high throughput, fairly long distances, and some ability to pass through walls. This spectrum, whether licensed or unlicensed, is the overwhelming choice of broadband network operators the world over and is apt to remain so for the foreseeable future.

Millimeter Microwave: Bandwidth at a Price

As we move past 10GHz, we enter what is known as the *millimeter microwave region*. In this region, multipath ceases to be a problem because the radiated energy can be tightly focused with a small, passive, dish-shaped antenna. A number of other constraints begin to manifest themselves, however, as well as a couple of singular advantages.

Generally, the shorter the wavelength, the more rapid the attenuation of the signal when it is propagated through the air, and in the region above 10GHz attenuation rises sharply from an initial level of 0.2 decibels (dB) per kilometer at 10GHz. To a certain degree, the ease of focusing millimeter microwave signals into narrow beams has the opposite effect because of the intense concentration of the RF energy within the beam, but still most network operators utilizing these frequencies do not attempt to transmit more than a mile.

Transmissions also become increasingly subject to atmospheric conditions, particularly rain. In fact, RF engineers have a term, *rain fade*, to describe the loss of transmission distance during periods of heavy precipitation. Rain fade can be addressed by increasing transmitting power, but in most places transmitting power is subject to regulation, and, in any case, RF amplifier power tends to go down with increasing frequency because of the inability of the power transistors used in the output stages of the transmitters to pass high-frequency waveforms that are also high in voltage. Indeed, solid-state amplifiers capable of developing even moderate power in the higher millimeter wave bands have been available only since the 1990s. Thus, signal attenuation in open air over distance becomes a substantial problem.

Attenuation of the signal in the regions above 10GHz is attributable to two causes: water vapor absorption and oxygen molecule absorption. Neither manifests a linear increase with frequency, but instead both exhibit wild fluctuations, with peaks of absorption followed by valleys and then further peaks, with an overall upward trend becoming evident. Incidentally, the patterns for oxygen and water vapor absorption are quite different, and their peaks and valleys do not coincide. Above 100GHz, oxygen molecule absorption quickly plunges to an insignificant level while the water vapor absorption trend moves mercilessly upward while still manifesting a series of high peaks and deep troughs as you go up in frequency.

From 10GHz to 30GHz, absorption of either sort is not a very serious problem, and only one absorption peak of any significance is present, that occurring at 23GHz. Consequently, the entire spectrum category is useful. Above 30GHz, water vapor absorption rises very steeply, exceeding 10dB per kilometer at 60GHz. Notwithstanding, spectrum has been allocated for broadband terrestrial use at 31GHz, 38GHz, and 39GHz, though water vapor attenuation is already quite severe above 38GHz. Figure 3-1 shows how frequency relates to atmospheric attenuation characteristics.

Another problem in the spectrum above 10GHz is the obstructing effect of not just walls but of even light foliage. Transmissions in these bands need absolutely clear line of sight, which obviously makes placement of base stations much more difficult.

A final problem, and it is a significant one, has been the equipment itself. The commercial microwave industry has been making reliable if expensive equipment for use in the bands below 25GHz for 30 years, but higher-frequency bands have been mainly used for radar until quite recently. Only in the late 1980s did bands beyond 25GHz begin to be exploited for commercial communications, the first use being in satellite systems where transmissions were highly asymmetrical and the high cost of millimeter wave equipment satellite was not of much significance in view of the already enormous cost of launch vehicles and the satellites themselves. Early attempts to build terrestrial equipment for these bands were not very successful, and only within the last year or so have millimeter wave data links reached a high degree of reliability. Figure 3-2 shows an example of a widely used base station radio for millimeter wave networks. Such equipment is typically more expensive than hardware intended for operation below 10GHz.

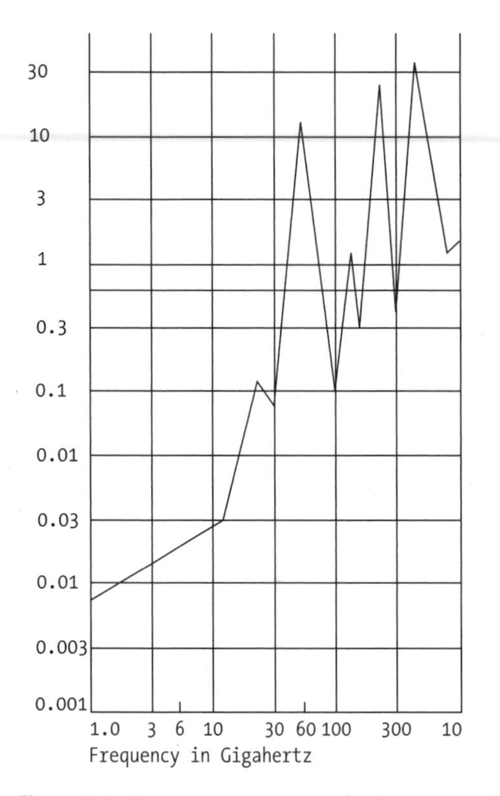

Figure 3-1. *Frequency vs. atmospheric attenuation characteristics*

Figure 3-2. *FibeAir 1500 subscriber terminal, courtesy of Ceragon Networks*

The cost of millimeter microwave equipment remains high today, though it has fallen a bit, and thus the equipment lends itself only to use with high-value customers such as mobile operators seeking high-speed backhaul and enterprises wanting medium-speed or high-speed packet services.

Nevertheless, the millimeter wave frequencies are not wholly problematic, and for a certain customer base they are actually advantageous.

Generous spectral allocations are the norm in the higher bands, so the operator has more inherent capacity with which to work. Now it is also true that bit-to-hertz ratios bear an inverse relationship to frequency and are usually only unity in the millimeter wave bands, but with bands spanning minimally several hundred megahertz and in some cases several gigahertz, the loss in spectral efficiency is more than offset by the additional spectrum. Add to that the fact that a millimeter wave airlink is generally much cheaper than a fiber link, and you begin to see a business plan.

The higher frequencies lend themselves to narrow beam transmissions, which is also an advantage in certain types of deployments. A tightly focused beam is ideal for a point-to-point connection where the full spectral allocation is assigned to a single user because that same spectrum can then be reused in another beam separated by only a few degrees with almost no interference between the two. And because high-frequency signals are subject to rapid attenuation, the spectrum can be reused in an adjacent cell whose center is as little as couple of kilometers away.

Submillimeter Microwave: Tending Toward Light

Beyond the SHF bands lies the extremely high frequency (EHF) region, a vast range of spectrum extending from 30GHz to 300GHz. (Acute readers will have realized by now that radio spectrum is arranged in "decades," where the uppermost limit of a region is always ten times the frequency of the lower limit.) Some 36GHz of this spectrum falls within the 802.16 standard. The region above 40GHz is somewhat inaccurately termed the *submillimeter microwave region* (in actuality wavelengths in the useful bands in this region are all above 1 millimeter) and has only recently become an option for the broadband wireless operator.

EHF is the last frontier of high-speed RF communications. Much of the spectrum has not been allocated by any government or standards body, and equipment manufacturers have not made many products available for this region. Still, I see significant opportunities in these bands, and I predict that activity there will increase considerably over time.

Currently in the United States two bands are in commercial use for high-speed data transmissions: an unlicensed band at 59GHz to 64GHz and a licensed band extending discontinuously from 71GHz to 95GHz that is known as the E band. The FCC is considering the allocation of still other bands to be located in the region above 90GHz. More equipment is currently manufactured for 59GHz–64GHz than any other (this is often referred to as the *60GHz band*).

Interestingly, as you have seen, the water vapor attenuation at 60GHz is extremely high, which would seem to make this spectrum a poor choice for outdoor airlinks, and in fact the band was originally allocated for indoor use. Outdoors, the practical limit for transmissions is less than 1,000 feet, though some operators see this as an advantage because it permits almost total frequency reuse in adjacent cells.

The band between 71GHz and 95GHz is situated within a deep attenuation trough with slightly more than 0.2dB total attenuation per kilometer at sea level—equivalent to that of the popular 38GHz band. Since much of that attenuation is because of oxygen absorption rather than water vapor absorption, attenuation drops precipitately at high elevations, and some authorities have suggested that the band would be well suited to trunk-line links connecting mountaintops or even links involving circling aircraft or stationary balloons. Until very recently this spectrum was only utilized on an experimental basis by Loea, a Hawaiian equipment manufacturer with deep expertise in radio photography. Loea owned a nationwide license and had plans to establish networks on a wide scale, though how quickly these can actually be developed remains uncertain. Recently the E band has become subject to expedited licensing by the FCC, which means that applicants can obtain licenses on a link-by-link basis, and can be fairly certain of obtaining licensing if no prior license has been established in the restricted geographical area irradiated by the narrow-beam point-to-point link.

All of these bands above 40GHz share one thing in common: allocations of bandwidth that enable them to achieve truly fiberlike throughput speeds. And yet, rather curiously, actual deployments remain quite uncommon. The largest I am aware of took place in Florida under the auspices of CAVU-eXpedient, a now defunct high-speed access provider that operated radio links at 38GHz and 60GHz and backed up the 60GHz links with free-air optical transceivers. Basing its business plan on a rapid rollout over several southern states, CAVU-eXpedient was unable to obtain third-round funding to continue its expansion and declared bankruptcy in 2002.

Nevertheless, I see considerable potential in EHF bands because of the relatively enormous throughputs they support. Services operating in the SHF spectrum may claim to offer fiberlike speeds, but only EHF services are truly capable of provisioning multigigabit pipes.

Free-Space Optics: Wireless Without the Radio

I mentioned earlier that EHF airlinks can rather easily be paralleled with free-space optical (FSO) transmission systems. Both FSO and EHF are vulnerable to adverse weather conditions—FSO to heavy fog and EHF to rain and snow—but, on the other hand, EHF is not much troubled by fog, and FSO is not obstructed by raindrops. Thus, used together, the two systems can achieve an availability figure that is at least an order of magnitude better than either alone over a given distance. And because the normal operating distances of the two technologies are similar and because both transmit tightly focused line-of-sight beams, the two work well in tandem.

At the same time, FSO may be considered to be a competitive technology with respect to 802.16 millimeter microwave. It addresses essentially the same market segments and shares a similar freedom from wireline infrastructure. It is also akin to millimeter microwave in that it is a young technology, appearing in commercial form in the 1990s and failing to establish a large market presence thus far.

In at least two other respects, however, the two access technologies are quite divergent.

FSO has the potential to offer significantly higher throughputs than EHF will ever achieve. Many of the same techniques used in fiber-optic networks, such as dense wave division multiplexing (DWDM), subcarrier multiplexing, 40GHz modulators, superfine optical filters, and optical Code-Division Multiple Access (CDMA), can also be used in FSO systems, and terabit speeds are theoretically possible over short distances. Indeed, some years ago, Lucent announced it had achieved 80 gigabits per second (Gbps) transmissions in the laboratory using DWDM alone. Surely that figure can be bettered in time. However, microwave throughputs are likely to achieve only incremental gains, and even these will require fundamental advances in high-speed devices and modulator circuits rather than building on technology that already exists, which is the case with FSO.

We must keep in mind, however, that terabit speeds or even large fractions thereof have not been demanded of any metro backbone except a lateral access connection extending from a metro hub. The great error of the telecom bubble was to assume that demand for bandwidth was insatiable and would drive the rapid adoption of faster and faster access technologies regardless of cost or other limitations. In fact, carriers have not found a great deal of demand for optical wavelength services in the range of 1Gbps to 10Gbps. It remains to be demonstrated whether demand would be more intense for a wireless service of comparable speed.

The other point of divergence between FSO and millimeter microwave has to do with distance. While some manufacturers of FSO equipment have claimed transmission distances of up to several miles, no commercial deployment yet shows that. Sending an infrared signal over great distances is not infeasible, but the power levels necessary to permit long-distance links render the transmitter unsafe to birds and to the eyes of any animal including Homo sapiens unlucky enough to blunder into the path of the beam. Higher power levels are also had at the price of throughput since laser modulators suffer the same trade-offs in terms of frequency and power level afflicting microwave output transistors.

To date, FSO systems have been costly, high maintenance, and critically short in useful range. Costs are coming down, and, at the same time, auto-alignment systems are reducing maintenance requirements substantially. Distance limitations will not be so easily solved, though, and these limitations will confine FSO to niche applications for the foreseeable future.

RF Orphans: The Low ISM Band and Ultrawideband

The first unlicensed frequency band was established by the FCC in the late 1980s, covered the range from 902MHz to 928MHz, and was dubbed the *industrial, scientific, and medical (ISM)* band. A friend of mine with a degree in RF engineering from Georgia Tech calls it the "dogs and cats" band because a host of rather incompatible devices occupy it, including garage-door openers, cordless phones, remote-controlled toys, and so on. Incidentally, the FCC has since authorized several additional ISM bands, and, the name notwithstanding, none has been much used in industrial, scientific, or medical settings.

Transmissions within this first ISM band penetrate walls with ease, and thus there are no line-of-sight issues with which to contend. On the other hand, the band is extremely crowded and has been almost since its inception, and it does not afford the network operator a great deal of bandwidth with which to work.

This band has been used by a number of wireless Internet service providers (WISPs), most notably Metricom, but is used relatively little today. However, a modest revival of interest in it appears to be under way. Transceivers are available from certain manufacturers, including Redline, Alvarion, and Trango. The best that one can say is that it represents an option, a means of eking out more bandwidth for a network or giving a network operator some wiggle room in a particularly interference-prone 2.4GHz environment.

Ultrawideband (UWB), touched on briefly earlier, is, in its pure form, a carrierless radio technology where the signal consists of a series of low-intensity pulses that themselves may span several gigahertz of spectrum and in theory encompass every frequency within that span. Recently the term has also been appropriated by an industry group touting a multiband Orthogonal Frequency Division Multiplexing (OFDM) technology that spans a considerable amount of spectrum but does in fact utilize carriers. UWB radios are supposed to coexist with radios occupying fixed bands, and, in theory, the UWB signal will appear as only a slight increase in background noise to a conventional receiver. The FCC has concluded that interference at levels needed for broadband access are in fact objectionable, though, and has limited the power of UWB equipment to a point where transmissions of much beyond 100 feet are impractical.

Originally conceived as a basis for ground- and foliage-penetrating radar systems, UWB has more recently figured in experimental stealth radios employed by the armed forces. Commercial prototypes intended for use in public access networks have shown a number of singular virtues including throughputs in the hundreds of megabits per second, near immunity to multipath distortion, and pronounced ability to penetrate buildings and dense foliage. Over-the-air UWB may ultimately play a prominent role in high-speed public communications networks, but it is going to have to clear some formidable regulatory hurdles before that happens. And given that powerful lobbies representing broadcast and telecommunications incumbents are vehemently opposed to it, those hurdles are not apt to be cleared any time soon, at least not in the United States.

Recently Southern California–based Nethercomm has developed a very interesting UWB system that could in fact play in metropolitan public networks. The Nethercomm system revives the old idea of waveguides whereby radio transmissions are sent through metal pipes that contain interference and permit senders to use as much bandwidth as they want. In this case the metal pipes are preexisting gas lines. Nethercomm claims that technology will support throughputs of several tens of gigahertz and will best any residential fiber-optic system. In theory, the notion should work, but we have *not* examined any actual implementations.

Licensed vs. Unlicensed Spectrum: The Operator's Dilemma

The licensing of radio spectrum has been in effect since the very dawn of commercial radio in the early 1920s, and the underlying notions informing such licensing are threefold. First, it sequesters generous allocations of spectrum for purely governmental use. Second, it further serves the government by deriving abundant revenues from the licensing process. And third, it prevents commercial users from interfering with one another by restricting each to a specific portion of the band.

Such considerations still exist today, though, at least in the United States, the second has assumed paramount importance.

Assuming that you can afford to pay for the license, licensed spectrum would appear to be a better medium in which to operate. The spectrum itself is a tangible asset with a real financial value, and the exclusivity of use provisions would seem to be a positive protection from interference. The issue is not so simple as appearances suggest, however, as the following sections illustrate.

The Unlicensed Frequencies: A Matter of Peaceful Coexistence

Beginning in the late 1980s a number of nations, including the United States, began to explore a new concept, sometimes known as *Open Spectrum*, that called into question the whole rationale behind the licensing of spectrum, or, in effect, making it the property of the license holder. The concept of Open Spectrum led more or less directly to the creation of the unlicensed bands.

This concept is that spectrum, rather than being private property, should be a **commons**—a shared resource available to all. To avoid the tragedy of the commons—that is, the mutual and simultaneous overexploitation of a scarce resource to the ultimate benefit of no one—the Open Spectrum will be open only to users of various kinds of spread spectrum radios. The users must have the following attributes:

- They have the ability to tolerate high values of interference.
- They are frequency agile within the band, distributing the signal among various coding sequences, subbands, and time slots so that no slice of spectrum is exclusively occupied by any individual user.
- They utilize various mechanisms for ensuring fairness such as mutual power control, network polling for controlling access, and the exchange of information among intelligent radios via control channels.

The first commercial spread spectrum radios appeared in the early 1990s at just about the same time that the FCC began to allocate unlicensed spectrum for miscellaneous uses. The first band so designated was the ISM band situated at 902MHz to 928MHz. This was shortly followed by the band extending from 2.4GHz to 2.4835GHz, the unlicensed band that is most subject to use today in transmitting data. This in turn was followed by two contiguous bands extending from 5.15GHz to 5.23GHz and another band between 5.725GHz and 5.825GHz. The 5GHz unlicensed frequencies are often referred to as the Unlicensed National Information Infrastructure (U-NII) bands and are subject to more usage restrictions than the lower frequencies, being explicitly designated for data transmissions only, not for the operation of remote control devices. In November 2003, the FCC allocated 250MHz of new spectrum extending from 5.470GHz to 5.725GHz, a band that is already in use elsewhere in the world. By this action the FCC has harmonized U.S. spectral allocations with those in many other nations, which should stimulate product development and bring down equipment prices.

Abroad, both 2.4GHz and 5.8GHz as well as 5.4GHz are commonly available as unlicensed spectrum, though not on a worldwide basis.

A further unlicensed band occupies spectrum between 24.0 and 24.25GHz in the United States, and another is situated between 59GHz and 64GHz. The 24GHz band is newly approved and has seen little or no use as yet. The 60GHz band has been commercially exploited on a small scale thus far, and the principal user is the now-defunct CAVU-eXpedient network in Florida. Rumor has it that the FCC will probably allocate additional spectrum for unlicensed usage in the midterm. Incidentally, the 60GHz band is in use in some other areas of the world as well.

The general rule for operation within the unlicensed bands in the United States is that users must be prepared to tolerate interference but must not generate undue interference themselves that would prevent others from utilizing the bands. All users are essentially enjoined to live and let live, and what is supposed to ensure that this will happen are rather severe output power limitations placed upon individual radios.

Since for many prospective broadband wireless operators licensed spectrum is simply unavailable, and unlicensed spectrum is the only option, such rather vague injunctions cannot but be disturbing. Won't interference inevitably increase as more and more entities use the bands? What is to prevent everyone from interfering with everyone else at a certain point, thereby rendering the bands useless and undermining the whole concept of Open Spectrum?

These are valid concerns, but in my experience of covering wireless broadband as a trade journalist, which dates from the earliest days of the industry, I have encountered relatively few instances where a network operator was severely hampered by high levels of interference. This is not to say that interference is not an issue; it is in fact a matter of some concern, particularly within the increasingly crowded 2.4GHz band. Still, for a variety of reasons, it is unlikely to be such as to cripple a broadband public network operating in the unlicensed bands.

Transmission Control Protocol/Internet Protocol (TCP/IP), the networking protocol specified for 802.16, is *best effort* in its basic form, which means that the nodes will retransmit in the presence of contention and dropped packets. If the traffic is suited to best effort, then the only result of a moderate increase in interference will be a slowing of throughput, which may or may not be noticeable to the user. If the particular application depends on maintaining a certain bit or error rate, however, then interference can pose a serious problem, and that would be the case for voice, multimedia, or real-time interactive applications. With the expected increase in such applications, concerns over interference and crowding in the unlicensed bands are likely to grow as well, though at the same time, new technologies for mitigating interference will be entering the marketplace.

The worst possible state of affairs is where users of unlicensed spectrum attempt to solve the problem by operating at maximum or above-maximum power levels in an attempt to rise above the interference. The situation then becomes akin to a rock band where every player is continually cranking up his amplifier in an effort to be heard above his band mates. The end result is a deafening cacophony where no one is heard distinctly. Unfortunately, inexperienced network operators often set their output levels at the maximum as a matter of course. The irony is that even when other operators are not present, the single operator may actually be degrading the signal at the receiver by increasing the level of self-interference from reflected signals.

Network operators faced with severe and intractable interference and with the resultant difficulty of honoring service-level agreements in its presence have a number of options. They can utilize high-gain, highly directional antennas that focus energy at the receiving node and establish a high carrier-to-interference ratio at that point in space. They can seek to identify interferers and see if a means of coexistence can be established through negotiation. They can seek injunctive relief by complaining to the regulating body if they can determine that an interferer or interferers are exceeding power limitations. Or they can try to confine their transmissions to less highly trafficked unlicensed bands, such as the 5GHz bands; however, by moving up in frequency to less-crowded regions of the spectrum, they may also be giving up effective distance and thus are somewhat restricting their potential markets.

In the future, smart radios with adaptive antennas may be able to distinguish interference from the desired signal and cancel it out at the antenna. This may not completely solve the problem because radio front ends may still be subject to overload if the amount of energy in the band rises to a certain level, but the intelligent radio will certainly provide an effective means of coping with interference in most cases. Currently, however, the cost of radios utilizing this technology is still high, and to date these techniques have been primarily employed in military radios and radar to thwart jamming and spoofing attempts.

In the future, you may also see mandatory coordination of all radios in an area utilizing unlicensed frequencies. Such coordination would serve to control the power levels of individual transceivers and maintain energy levels below certain stated limits. Power control of this type has long been a characteristic of licensed CDMA mobile phone networks and has been proven to work well. Extending it over a motley assemblage of unaffiliated users may pose some policy problems, though.

Licensed Spectrum

Within the lower microwave region, licensed spectrum for broadband public services is relatively scanty, and only two significant allocations exist. The first, called Wireless Communications Service (WCS), is located at 2.3GHz and has only 30MHz of bandwidth. The second, called Multichannel Multipoint Distribution System (MMDS), is located in the 2.5GHz to 2.7GHz region and provides the operator with almost 200MHz of spectrum.

The WCS spectrum was originally allocated to AT&T on an experimental basis, and the latter set up the Project Angel program to exploit it. Another license was issued to the now-defunct Metricom that used these frequencies in conjunction with unlicensed spectrum to provide Internet access. Originally, Project Angel was intended to support a fixed wireless telephone system, an arrangement often referred to as *wireless local loop*, but as the experiment progressed the network evolved into a high-speed access service. AT&T eventually abandoned the project, and licenses were made available to other operators.

The MMDS spectrum is generally considered the most desirable licensed spectrum in the lower microwave region, at least in the United States. Originally intended for one-way "wireless cable" television programming distribution, the MMDS spectrum was sold to hundreds of licensees across the United States, some with strong regional presences extending across scores of metropolitan markets and some confined to a single market. With few exceptions, MMDS television distribution systems have failed, and most of the local licenses have been sold to a handful of telecommunications giants, chief among them Sprint and BellSouth.

I will dispense with any detailed history of attempts to utilize this spectrum for data, attempts that began in the late 1990s on an experimental basis and were seriously under way by the year 2000. One such deployment, that of Sprint, managed to attract some 90,000 subscribers across a number of metropolitan markets, but that company has not taken on new subscribers for more than a year and has given little indication of its plans for this spectrum in the future. Rumor has it that Sprint simply did not understand how to properly market line-of-sight RF technology, which resulted in huge losses through rolling trucks to dissatisfied customers and proving to them that they really could not receive a signal because of some obstruction between their location and the base station. Others have suggested that the company could not operate the network cost effectively with first-generation equipment and ceased expansion to await the arrival of better radios. Other progenitors, such as BellSouth and MCI, proceeded more cautiously, setting up what were in effect merely pilot programs. No one to date can be said to have achieved a solid success in the MMDS bands within the United States.

The fairly stringent line-of-sight requirement of first-generation equipment was a major impediment for the pioneers, one that put them at serious competitive disadvantage with cable and digital subscriber line (DSL) service providers. But a further serious problem was presented by the distribution of spectrum within the MMDS bands.

The MMDS spectrum is divided into thirty-three 6MHz channels, 6MHz being the bandwidth of analog broadcast television channels according to the National Television Standards Committee (NTSC) standard in force since 1941. Interspersed among the MMDS channels are Instructional Television Fixed Service (ITFS) television channels allocated to nonprofit institutions such as churches and schools and used for educational and instructional television. The presence of these intervening channels severely hampered the MMDS system operator in a number of ways. It placed significant constraints on the design of the radio by requiring tight filters to keep out interference from neighboring ITFS channels and to mitigate interference from the MMDS transmissions themselves and severely limited the flexibility of the network operator in assigning spectrum to users or in the choice of a modulation system.

Some MMDS operators leased channels from ITFS licensees to control a block of contiguous spectrum, but that option was not always available, and the presence of multiple operators within the MMDS bands made for a nearly untenable situation.

In 2004 the FCC reallocated the spectrum in 2.5GHz to 2.7GHz range to permit larger blocks of contiguous spectrum, which has resulted in a vast improvement. Whether it will lead to the creation of successful services in these bands remains to be seen.

MMDS spectrum is not widely available to independent operators in the United States, but it can be leased or purchased in certain communities, generally in rural areas where the licenses are still held by failing MMDS television system operators. I should also point out that by aggregating spectrum in MMDS and the 2.4GHz ISM band, an operator can command approximately 260MHz, a very considerable swath of bandwidth. I think that MMDS spectrum, where obtainable at a reasonable price, is potentially very valuable. Incidentally, MMDS spectrum is available in many other places in the world, generally at around the same frequencies.

One further note: Several years ago a company calling itself Clearwire leased large amounts of ITFS spectrum in various markets throughout the United States. In 2004 Clearwire dissolved and transferred its leases to a new company called Clearwire Ltd., which is owned by former cellular telephone magnate Craig McCaw, who subsequently leased and purchased additional spectrum in the United States and Canada. The new Clearwire marks the largest independent effort to establish broadband wireless services in the United States, and the industry as a whole is watching it closely.

The next licensed bands of importance reside at 18GHz and 23GHz. Both bands are designated for point-to-point connections and provide roughly 100MHz of bandwidth. Local licenses for the band located at 23GHz are readily obtainable and are low cost. To date, 23GHz has been used primarily for high-capacity wireless bridges for enterprise networks and has been little used by service providers. Both bands are well suited to providing fairly high capacity backhaul services, and both operate strictly line of sight. Ranges of up to several miles are possible with suitable antennas.

A pair of 400MHz wide bands is located just above 24GHz adjacent to an unlicensed band mentioned in the previous section. The first band extends from 24.25GHz to 24.45GHz, and the second runs from 25.05 to 25.25GHz. This licensed spectrum, which is designated Digital Electronic Message Service (DEMS) by the FCC, became the almost exclusive domain of Teligent, a company that went bankrupt in 2001 and then reorganized. Most of the spectrum was subsequently acquired by First Avenue Networks, a Virginia-based independent carrier. Spectrum in this region is also available in a number of Latin American countries.

The 24GHz-licensed band is designated for point-to-multipoint, as well as point-to-point use, and is strictly line of sight. Transmission distances are somewhat greater than for higher millimeter microwave frequencies.

Spectrum at 26GHz is widely available for point-to-point transmissions in Europe and Asia but not in the United States.

Several bands of frequencies between 28GHz and 31GHz have been designated for point-to-multipoint services known as Local Multipoint Distribution Service (LMDS). Frequency divisions are as follows: 27.5GHz to 28.35GHz, 29.1GHz to 29.25GHz, 31.0GHz to 31.75GHz, 31.75GHz to 31.225GHz, and 31.225GHz to 31.300GHz. Most of the spectrum in these bands in the United States was purchased in FCC auctions some years ago by a few large corporate entities, including XO Communications and Winstar. Spectrum at 28GHz is commonly used in

Latin America and Europe, as well as in North America, though exact frequency allocations differ from country to country.

LMDS spectrum is only spottily available today in the marketplace in the United States. Some scattered independents in secondary and tertiary markets own such spectrum and may be induced to sell it in some instances, but the largest players, though all bankrupt now, disposed of their holdings in blocs, or else were rescued and refinanced. LMDS bands have the capacity to support a great deal of traffic, but they are strictly line of sight, and the links are undependable beyond a couple of kilometers. I think LMDS can provide the basis of profitable services in selected markets, but I am skeptical as to the viability of point-to-multipoint architectures at these frequencies. Provisioning very fat point-to-point pipes to high-value customers and using these frequencies for cellular backhaul probably makes more sense.

The next licensed frequency of interest is located in two bands in the 38GHz region and provides more than 500MHz of bandwidth in the United States and in Canada. The 38GHz frequency is widely used in Latin America as well as in the United States, but frequency allocations are quite different there from those in North America. This is a region of deep water vapor attenuation, and ranges are even shorter than is the case for LMDS. To date this spectrum has been used for point-to-point and point-to-consecutive point services. ART and Winstar purchased great numbers of local licenses several years ago, but neither established a nationwide footprint, and both in fact went bankrupt. Some 38GHz licenses are available today as distressed assets. Most of the strictures applying to LMDS also apply to 38GHz except that the useful range is considerably less.

Another licensed band is available at 39GHz, and for all practical purposes it is similar to the 38GHz band. First Avenue (`http://www.firstavenet.com/`) has bought a great deal of this spectrum and leases it, primarily for backhaul, to other carriers. A further band extending from 75GHz to 85GHz has been allocated to a single carrier, Loea, on an experimental basis. Recently the FCC has allocated additional licensed spectrum in the region extending from 70GHz to 90GHz and has made it available on an expedited basis, as has been the case with 18GHz and 23GHz.

Different Uses for Different Frequencies

Frequency and bandwidth are major factors affecting the scope of the services that the network can deliver and the customers who can be reached. The band one selects for one's airlink has a considerable bearing on the market one can address.

Lower Microwave: Primarily a Residential and Small Business Play

The lower microwave frequencies, are, as you have seen, governed by the 802.16a standard that actually extends to 11Ghz. Still, in my estimation, typical lower microwave deployments are best not attempted at frequencies much above 6GHz, and in the United States no commercial spectrum is available between 6GHz and 18GHz that is authorized for broadband access applications. Spectral allocations of course vary from one nation to another, but generally the region between 6GHz and 10GHz has been little used for providing broadband access anywhere in the world.

Although spectral allocations in the lower microwave region vary from one nation to another, most bands are under 200MHz in width and often under 100MHz. Moreover, most antenna systems designed for use in these regions simply will not allow the entire bandwidth to be reused within a few degrees of arc, so network operators must allocate their assigned spectrum—be it 30MHz, 50MHz, 100MHz, or 200MHz—to a number of contiguous users who will all be encompassed within the same transmission beam and thus will all be presented with the full allocated spectrum. Coincidentally, those same users will also be exposed to every transmission occurring over that spectrum within the sector defined by the antenna beam.

The practical implications are numerous.

First, users are going to get a fairly small allocation of spectrum for their own use, perhaps no more than a few hundred kilohertz in a fully subscribed network. Second, for reasons of security, the network operator cannot depend entirely on channelization or modulation coding to segregate transmissions from one another and to ensure privacy. Instead, further privacy measures are advisable, preferably ones that entail the encryption of each transmission and the installation of an encryption key management system to make the encryption process both transparent to the user and extremely difficult to penetrate on the part of interlopers. Chapters 6, 8, and 9 discuss such security and encryption methods at length.

The relatively small amounts of bandwidth that can be accorded individual users in the lower microwave bands restrict the network operator's ability to court large enterprise customers, and the basic service offerings end up competing with DSL and cable services for the small business customer and the small office home office (SOHO) customer. Residential customers for high-speed access may also be served, but there the wireless operator is apt to enjoy no particular advantage in markets where DSL and cable are already well entrenched.

Addressing the Bandwidth Problem in the Lower Microwave Regions

It should be noted that some radios do permit simultaneous operation in two different bands, significantly increasing the amount of bandwidth available to the network operator and ultimately to the subscriber. Most commonly, dual-band radios will use one band for the uplink and another for the downlink. One may, for instance, use the 2.4GHz unlicensed band for an uplink and the MMDS licensed bands (2.5GHz to 2.7GHz) for a downlink. Downlinks, it should be understood, normally utilize far more bandwidth than uplinks; in other words, the network is asymmetrical—the theory being that most residential and small business users tend to download far larger files than they transmit, and that the Web browsing experience will be greatly enhanced by a faster downlink as well.

It is possible to design radios that can span more than two bands; one could, for example, build a radio that could operate in the lower ISM band (902MHz to 928MHz), the unlicensed band centered at 2.4GHz, and all three of the three unlicensed bands between 5GHz and 6GHz. A company called Wireless Inc. put out just such a product a few years ago but was unable to find a market for it and ceased production. The basic concept remains valid, however, and I predict that as demands for bandwidth grow among 802.16a network operators, it will be revived.

In the midterm, radios will begin to appear that are not bound to fixed bands at all and will enable the network to use almost any frequency desired; that is, the radio would not have to be factory tuned to just a few bands but would have the flexibility to select any band and could automatically accommodate itself to changes in bandwidth allocations on the part of governing bodies. Termed *frequency-agile* radios or *software-defined* radios, such products will in most cases be able to select a modulation system as well.

Software-defined radios exist today, and at least a score of companies are active in this area, but no one is making a WiMAX-certified product conforming to the 802.16a standard, and in fact most sales to date have been to military organizations. The majority of software-defined radios made thus far have utilized costly field programmable gate arrays (FPGAs) and have not met the price requirements of network operators. Incidentally, several have combined frequency agility with an adaptive array antenna using the extreme processing power provided by the FPGAs to perform multiple functions.

Frequency agility is unquestionably desirable in governmental applications. It provides a radio operator with a means of concealing a transmission and evading jamming, and it allows safety personnel to contact all relevant agencies over whatever frequencies have been allotted to them. But in a commercial setting where relatively few bands can be legally occupied by any individual network operator, the need for such flexibility is unclear at present. Many advocates of software-defined radios look forward to a day when radio spectrum is brokered and made available to consortia of network operators as needed, with payments automatically being made to license holders or with bandwidth being swapped among participating networks within a peering relationship. Such a grand schema would certainly result in much more efficient use of spectrum than is presently the case, but it is difficult to envision how multiple networks with countless users could all coordinate their transmissions so as to avoid interference. Everyone would have to be equipped with an intelligent radio, and each radio would have to connect to an overarching computing grid devoted to the management of all available spectrum.

I will not speculate further as to how or even if such a virtual organism could come into being, and I do not see even the beginnings of such a regime occurring within this decade. Nonetheless, some degree of frequency agility will manifest itself in the lower microwave regions within five years, and most likely it will involve subscriber terminals utilizing 802.11, 802.16, and 3G mobile networks as the situation warrants and in accordance with roaming agreements among a relatively few incumbent carriers.

Software-defined and frequency-agile radios will undoubtedly influence both spectrum use and the posture of regulatory bodies over time, and they will greatly alter the situation of network operators by making significantly more spectrum available to them, perhaps in amounts approaching 1 gigahertz. Eventually the increase in available bandwidth will expand the uses to which the network may be put and will open the possibility of the network operator offering a wide range of high-fidelity multimedia content as well as simple high-speed access and LAN extensions. At the same time, I caution broadband wireless operators today not to predicate their business plans upon the imminent arrival of such technology in the marketplace. That may be years away, and in the meantime operators will have to adapt their business to relatively scanty bandwidth.

I will now add a final word on current throughput constraints in the lower microwave regions and how they may be overcome in the future by technologies other than software-defined radios and the related adaptive array antennas. Over the course of the next several years, radios will steadily increase their ability to resolve signals in the face of greater and greater amounts of noise and interference and in their ability to reconstruct information that has dropped below the noise floor. These abilities will allow operators to pack more and more information into a signal by increasing the number of discrete phase and amplitude states used to convey information and, in the case where subcarriers are employed, decreasing the spacing between them. All of this will lead to higher throughput and, combined with the aforementioned technologies of adaptive antennas and software-defined radio, may eventually allow the achievement of fiberlike speeds within the lower microwave region. Fiber itself will not stand still, however, though it remains to be seen what applications will emerge that will require throughputs in the tens of gigabits per individual users.

In any case, the midterm future of wireless broadband promises to be an era of very abundant throughputs and greatly expanded service opportunities for the operators. Five-year market projections should consider the likelihood of fundamental improvements in the core technology and should not rest upon the assumption that the business will remain much as it is today. The operator should take it almost as a given that generations of equipment will be short and that capital improvement will have to be undertaken on an ongoing basis in order to remain competitive. For the foreseeable future, the low microwave wireless broadband network is not going to be a set-it-up-and-forget-it cash cow.

In the current regime characterized by throughput rates that are generally competitive with those of cable and DSL, and by network capacities that are generally much lower, network operators must proceed cautiously, however, recognizing that they lack the resources to be all things to all customers. Their aim must be to utilize the capacity of the network as efficiently as possible and target those customers who will bring them the best return on the infrastructure investment.

This aim cannot be achieved in most markets simply by presenting subscribers with a "pipe," or raw capacity. Raw capacity to support broadband access is becoming increasingly available and increasingly commoditized, and the mere fact that a wireless carrier is offering capacity is scant inducement for the customer to subscribe. Wireless broadband represents an opportunity for competitive access provider, perhaps the only opportunity to enter many markets, but is not inherently highly attractive to end users simply by virtue of being wireless. If the wireless operator is competing with well-entrenched wireline incumbents who are prepared to wage price wars, and if that same wireless operator is offering nothing more than basic services, his prospects of succeeding are poor because the service offering is just another broadband choice and one that is in certain respects technically disadvantaged vis-à-vis the others.

Chapter 2 mentioned the major value-added services that can be supported over wireless networks. The following sections attempt to be more specific with respect to the lower microwave region.

Looking at the Range of Services

No value-added broadband service exists that cannot be supported by a lower microwave broadband wireless network, but because some services are heavily bandwidth intensive and some are not, the operator must decide on the mix of services to be made available to subscribers and the marketing emphasis that should be brought to each.

Basic Access

Currently, basic Internet access will constitute the core service offering just as is the case with DSL and cable. The operator may want to serve as the Internet service provider (ISP) or may want to establish a relationship with one or more third-party providers. For an independent, competing with a giant such as America Online or EarthLink in terms of content and programming is well nigh impossible, and, in any case, customers may specifically want to have a high-speed link to their existing ISP. If that is the situation, working with incumbents makes good sense.

VPN and Transparent LAN

Virtual private networks (VPNs) and transparent LANs, also known as LAN *extensions*, are the most popular business-related services today and pose interesting possibilities for the network operator inasmuch as they do not need a lot of additional bandwidth beyond that required for simple high-speed access. They do involve additional investment in infrastructure and some fairly hard decisions on the part of the network operators based on their perceptions as to how the evolution of business data services is likely to proceed.

A VPN is a secure communication from a remote location back to a corporate LAN that often involves encryption. A transparent LAN service takes the concept a step further by providing the remote user the means to run all LAN applications with similar ease at the remote location as at the corporate headquarters and not simply to access corporate databases.

VPNs and LAN extension can be accomplished by various means. The older method is to utilize tunneling protocols and special hardware for performing the necessary encryption quickly. The preferred method today is to utilize a carrier-grade IP router or Ethernet switch with the built-in capability to set up Ethernet or IP VPNs. Such devices will exhibit a high degree of automation and scalability and low network overhead.

Most edge routers sold today—that is, routers designed for use in a metro network—support IP VPNs through the agency of the Multiprotocol Label Switching (MPLS) ancillary protocol. Manufacturers of the latter include Cisco, Juniper, Laurel Networks, Riverstone Networks, TiMetra, and Vivace. Most of these products also include other functions such as Ethernet switching and ATM switching, though the precise mix of functionalities varies from one platform to another.

Enhanced Ethernet switches form another possibility, specifically those designated as "metro Ethernet" and designed for the carrier marketplace. Atrica and Extreme Networks are the principal manufacturers in this category. But inasmuch as 802.16a networking products are based on IP rather than Ethernet, and because IP/MPLS has a much greater capability for setting various service levels to accommodate different types of customers, an edge router represents a better choice for enabling VPNs.

Voice Telephony

Another fairly obvious service offering beyond simple high-speed access is voice telephony. Either circuit voice or packet voice may be supported on 802.16a platforms, but circuit voice is quite bandwidth intensive, requiring the reservation of a 56 kilobit per second (Kbps) channel for each voice transmission. In contrast, IP telephony vocoders (digital speech compression devices) can operate at rates as low as 2Kbps with good fidelity. Fifty-six kilobits may not sound like much, but it takes only 24 voice circuits to eat up 1.5Mbps of throughput. If the total throughput of the available band is only 100Mbps, then a couple of thousand voice circuits could take up all the resources of the network.

Another factor to consider when choosing between circuit voice and IP voice is the cost of the infrastructure. Circuit telephony is delivered within the metro via class 5 telephone switches costing millions of dollars. IP telephony, on the other hand, is supported by *softswitches,* which are software emulations of circuit switching run over packet networks and utilizing large, hardened, general-purpose workstations such as the Sun Netra platform. Total cost is in the tens of thousands of dollars. Among the newer, specialized manufacturers, Sonus, Tacqua, Syndeo, and Oresis are prominent, and old-line equipment manufacturers that have added softswitches to their product lines include, most visibly, Nortel, Telcordia, and Marconi. Incidentally, hybrid softswitch/circuit switches exist, of which Nortel's product has been the most successful. It should be noted that within the last two years VoIP architectures have been developed that can dispense with softswitches but only in cases where a pure IP network has been established between the two calling points. In cases where the call must traverse the public switched telephone network (PSTN), a softswitch will still be required.

IP telephony may rightly be considered an emergent technology, one characterized by warring standards bodies and a lack of consistent, generally accepted approaches to network engineering. Some systems, for instance, use devices called *gatekeepers* that perform access control and address translation while others dispense with them. Conversely, some incorporate the switching and PSTN interface functions in a single box, and others favor a distributed platform. Another problem has been lack of real functional equivalence between traditional class 5 switches and IP softswitches. IP softswitches to date have not supported the entire myriad of features programmed into circuit switches, which in sum represent millions of lines of code.

Another problem with IP telephony is the requirement for special customer premises equipment that will interface with standard desksets. If the subscriber is a business and already has an IP private branch exchange (PBX), that is not a problem, but otherwise voice will have to be recovered from the packet stream coming into the subscriber terminal and subsequently sent to the desksets via ordinary copper telephone lines. That often involves a separate black box, though some broadband wireless access equipment does support IP telephony today.

As matters stand today, the wireless broadband operator is unlikely to be able to compete with the wireline incumbent as primary providers of voice services whether they offer circuit or packet voice services. Still, if the wireless operator is in a position to provide a subscriber expeditiously with extra phone lines, and the DSL or cable operator in the area cannot do so (or, for that matter, the telco incumbent), then the wireless operator enjoys an advantage. And, if the opposite is true—that is, if the competitive broadband access providers are offering second-line telephony—then it is incumbent upon the wireless service provider to match those service offerings.

In the interest of broadening the service portfolio, the wireless operator should definitely consider offering long-distance voice service as well as local service. In this instance the operator will inevitably be functioning as a reseller, and the long-distance service itself may be a loss leader since the operator can compete only on the basis of price and can offer no unique calling features. One may question the wisdom of offering a service that is not in itself highly profitable, but studies have shown that operators providing the greatest number of converged services are likely to be the least affected by "churn," the tendency for subscribers to switch service providers impulsively.

Long-distance telephone services may be either traditional circuit, generally carried over ATM, or IP. If IP is chosen, make sure that specified peering arrangements are in place within end-to-end connections between both calling parties. Such peering arrangements involve the carriers providing long-distance data transport, and in the case of voice they must ensure low latency and jitter and minimum transit times. Ordinary best-effort service through the public Internet will not do. Chapter 6 covers this topic in more detail.

Mobile Voice

Finally you must consider the issue of broadband operators moving into the mobile telecommunications space and competing with cellular and Personal Communications Service (PCS) operators. Several hardware manufacturers, including IP Wireless, Flarion, MeshNetworks, and Navini, already support mobility, and each has suggested that mobile voice services are well within the scope of a broadband wireless operator.

Such services could enhance the attractiveness of the broadband wireless network to some customers especially if a single terminal, presumably some sort of smart phone, could be made to serve as both a phone and a high-speed access device—which is really the whole notion behind third-generation cellular. But before this kind of far-reaching convergence can take place and win the acceptance of network operators, several developments will have to happen.

First, the 802.20 mobile broadband standard will have to be completed (or, alternately, the less comprehensive 802.16e amendment to the 802.16 standard will have to be completed). To what extent this will be complementary to 802.16a remains to be determined, but it will most likely apply to the same bands currently covered by 802.16. Second, the handset manufacturers will have to produce phones that conform to the new standard, because without phones, mobile telephony over broadband will remain conjectural. Third, some sort of modus vivendi will have to be fashioned by the broadband wireless service providers and the mobile telecom incumbents that themselves are moving toward fourth-generation systems, which will likely have much in common with the 802.20 networks. Already mobile phone operators are discussing using unlicensed frequencies in the 2.4GHz band to augment the mobile bands during large downloads and are anticipating further allocations of licensed bandwidth to support true high-speed services.

I do not expect 802.20 to be finalized until 2006 at the earliest, and there is no telling when 802.16e will emerge. A further span of time will be required for chipmakers to embody the standards in silicon, and yet more time must elapse before any kind 802.20- or 802.16e-compliant terminals become available. Gaining the support of handset manufacturers becomes essentially a chicken-and-egg conundrum since the manufacturers are not interested in undertaking a small production run. Still, Nextel (now part of Sprint), with its proprietary Integrated Digital Enhanced Network (IDEN) system, managed to obtain handsets for its few million subscribers, so 802.20- or 802.16e-enabled phones are not inconceivable within two or three years, though prices are unlikely to be comparable to those of cellular units.

In the meantime, the cellular industry will proceed slowly but inexorably with the development of fourth-generation systems. Fourth-generation will share in common with 802.20 the use of packet transmissions to convey voice messages and an elimination of circuits altogether except to serve legacy customers. It will also provide much higher throughputs for data, purportedly in the megabits per second, at least in burst mode.

The question then becomes, how will the two services differentiate themselves in the marketplace, or will they? Similar questions were asked concerning PCS when it emerged in the early 1990s, the supposition at the time being that PCS would fill a different market niche. It did not, and ultimately fourth-generation cellular and broadband wireless may not fill different niches either.

Although mobile voice may be a desirable service offering for broadband wireless operators in that it may be something to entice customers onto the network, it may also be a loss leader. Mobile voice is already a price-eroded business, and the arrival of new service providers and one-number portability cannot improve that situation. For now, fixed voice is the proper concern of broadband wireless operators, however, and they will have plenty of tasks with which to occupy themselves as they attempt to work through the intricacies of implementing IP voice on a grand scale. Currently, only IP Wireless manufactures broadband wireless equipment for mobile applications. Figure 3-3 shows a Compaq personal digital assistant (PDA).

Figure 3-3. *Compaq PDA with wireless IP PCMCIA card, courtesy of IP Wireless*

Conferencing

Video- and audioconferencing are well within the capabilities of 802.16a systems and are services the network operator should strongly consider promoting, especially videoconferencing. In the past, very expensive proprietary hardware/software platforms were necessary to set up a videoconference with acceptable image quality, but today a multitude of high-performance IP videoconferencing software products are available and can be run over ordinary edge routers. Videoconferencing, while never experiencing the explosive growth predicted for it in the past, must be considered a successful application and is increasingly utilized by enterprises for training purposes and for controlling travel expenses. Furthermore, it has always been a high-value application for which businesses have been willing to pay premium prices. Finally—and this is good news for the broadband wireless service providers striving to differentiate themselves from other broadband providers in the local marketplace—it has seldom been offered by anyone other than incumbent telcos or large long-distance service providers.

Telemetry

Wireless telemetry, discussed in brief in the previous chapter, is a business application for which there is a considerable sales potential and which, at the same time, makes small demands upon bandwidth. To date, few broadband service providers have gone after the telemetry or remote monitoring market, and, in many cases, telemetry is done over private networks, often using special narrow telemetry bands in the UHF/VHF region. Wireless networks are particularly well suited to telemetry functions, and the wireless broadband operator can often pursue telemetry accounts without worrying about competition from the wireline broadband providers. Marketing telemetry services is not easy, however. There is no consistent profile for the users of such services and no established venue for advertising them. Probably the best strategy in selling such services is to bundle them in with other data services and let prospective business customers know that the wireless network can support telemetry.

Backhaul

The lower microwave network operator can also function as a carrier's carrier by offering backhaul to other types of network operators such as mobile telephone operators or Wi-Fi *hotspot* operators. This can be a fairly lucrative business and is one that may enable the wireless broadband carrier to avoid paying for tower space by utilizing the facilities of the carrier being served.

A backhaul link is quite distinct in terms of network architecture from the sort of point-to-multipoint topology that tends to characterize last-mile access networks. Instead of apportioning the available spectrum to a number of simultaneous users and employing medium-beam-width antennas to illuminate a designated sector, the operator uses a single, narrow-beam point-to-point airlink often described as a *wireless bridge*. This will normally use the entire bandwidth allotted to the network. With 5.8GHz spectrum, which is often employed for the purposes of backhaul, 100Mbps throughputs are easily attainable over distances of several miles.

Storage Area Networks

Yet another business application that conceivably could be supported by wireless is storage area networking, also briefly touched upon in Chapter 2. My assumption is that the higher microwave bands encompassed by 802.16 are better suited for this purpose, however.

Data storage is a network application where vital information is off-loaded to remote storage facilities and invoked thereafter as it is needed. Storage equipment sales, including those for networking gear, have remained brisk when other areas of business computing and data networking have been plagued with declining demand, but thus far service providers have had indifferent success in selling storage networking services. Most storage still takes place within a LAN environment and involves special network protocols such as Fibre Channel and ESCON. Fairly recently IP and Ethernet storage solutions have come on the market, and many industry analysts think these will displace the old closed storage standards eventually. To my knowledge, storage services have never been offered over a broadband wireless network, and no 802.16 equipment is currently architected to provide such services. However, storage services may be offered by wireless operators in the future and constitute a potential market worthy of examination.

Figure 3-4 provides an example of a popular storage network element, the Sanrad Ethernet storage switch.

Figure 3-4. *The Sanrad Ethernet storage switch. Sanrad is a major manufacturer of Ethernet-based storage network products. (Courtesy of Sanrad)*

Entertainment Services

Broadband wireless, at least in the United States, began as a purveyor of entertainment fare, principally video programming, and at least one broadband wireless commercial service, Digital Video Broadcast (DVB), is still striving to compete with cable and direct satellite in this market, though scarcely at all in North America as yet. I wish them luck, but the prior history of broadband wireless operators in this marketplace is not encouraging.

For the lower microwave operator to attempt to provide the same type of video programming emanating from the cable networks and satellite broadcasters is generally a mistake. Even with digital compression, the network will have difficulty supporting even 100 channels in most instances and is likely to have to content itself with far fewer. Remember, cable operators have potentially several gigahertz with which to work, of which most systems today utilize at least 750MHz. The broadband wireless operator is lucky to have 200MHz and is more likely to have 100MHz or less. If the network operator is committed to the notion of converged services, which is probably the only business model that will succeed in the long run, then that 100MHz or less must be divided among several other services—basic access, VPNs, LAN extension, telephony, conferencing, telemetry, and so on—and the decision to do video will leave little bandwidth for anything else. To date, few subscribers to broadband services have signed on primarily to get video, and very high bit rate DSL (VDSL), the one new two-way access technology specifically designed to support video, has enjoyed at best limited success in the marketplace.

At the same time, VDSL could provide a model for a video service with the potential for success, one that would avoid the inherent constraints in the old MMDS model.

All broadband wireless video networks to date have resembled cable networks to the extent that all channels were sent to all viewers simultaneously in just the same manner as traditional over-the-air UHF and VHF television broadcasts, and, in a strictly one-way network where the viewer cannot signal back to the studio, that is practically the only way to distribute the content. But in a two-way network where each user is assigned some portion of the spectrum for exclusive use, then the viewer need not tune up and down the spectrum at the premises in order to select a program, and the program itself need not be assigned to any particular channel. Instead, the tuning function could occur at the base station, and the selected program would then be sent down a channel assigned to the subscriber rather than to the program itself.

In a sense this constitutes a form of video on demand (VOD) inasmuch as an individual program is being sent to an individual subscriber, and indeed such a system could support VOD in the strictest sense where a viewer selects from a library of prerecorded materials stored on a video server. But it would also support the transmission of regularly scheduled programming that would play continuously in studio and would then be accessed by individual subscriber tuners located not on the subscriber premises but in the studio itself.

This discussion focuses on the mechanics of distribution and not the full business case for doing video, and it must be noted that video incumbents, chiefly the cable operators and secondarily the satellite broadcasters, enjoy advantages other than simply possessing very wideband distribution pipes. Equally important are the mature relationships they enjoy with *content producers,* a term that is generally synonymous with the studios. Major studios are increasingly picky as to whom they choose to distribute their wares, in large part because of the mounting incidence of piracy and unauthorized distribution, and independent broadband operators have difficulty forging the same close relationships that the cable operators have maintained for decades and the satellite operators have had for a single decade. Independents can obtain popular content in a reseller arrangement through mPhase, a video-delivery platform developer serving independent rural telcos seeking to get into the television business with video over DSL, but they still have to meet certain criteria relating to network security, and they are apt to pay considerably more than the big cable multiple service operators (MSOs) do for the same content relative to the number of subscribers served. Another source of content is the National Cable and Telecommunications Association (NCTA), a trade organization for independent telcos and cable companies.

In addition to the more familiar video programming consisting of episodic television programming and feature films, many newer types of entertainment service offerings exist, including multiplayer gaming, music on demand, Internet radio, and interactive television shows. All of these have excited the interest of the telecom carrier community to varying degrees, but none has proved to be a highly profitable business as yet. Wireless broadband service providers should monitor the progress of new entertainment service offerings in the marketplace and should strive to achieve first-mover advantages when such offerings appear to be market ready. At the same time they should be cautious about entering a market too soon or embracing highly speculative formats for which there is yet no demonstrated subscriber demand.

Wi-Fi Hotspot Services

Hotspots are essentially public local area networks where connectivity is provided to transient users, generally via 802.11-based wireless LAN interfaces. Wi-Fi hotspots are located in such venues as airports, shopping malls, convention centers, hotels catering to business travelers, coffee shops, truck stops, and so forth, and they form a convenient means for the traveler to check e-mail, surf the Web, or consult a database. Compared to other access points, hotspots are cheap to install, and a multitude of service providers large and small have entered the hotspot business and have begun to build an infrastructure of access points. The hotspot business is at a fairly early stage of development but is growing rapidly, and the hotspot phenomenon is now international in scope.

Hotspot services themselves are not cost effective when delivered with current 802.16 equipment, but 802.16 could prove useful in backhauling hotspots to Internet access points, and 802.16 operators could either set up hotspots on their own or approach existing operators to sell them backhaul. I do not believe that hotspot services form a sufficiently mature business for the broadband operator to embrace as a primary service offering, but hotspot access could be presented to subscribers to residential or SOHO services as a value add. If the network operator decides to offer such a service, a specialized billing plan must be put in place, and roaming arrangements must be established so that subscribers to other networks will be able to utilize the hotspots set up by the 802.16 network operator.

Chapter 5 covers such requirements at length.

Higher Microwave: Abundant Throughput Speed but Fewer Applications

The higher microwave frequencies included under the original 802.16 standard are best discussed in terms of the services they cannot readily enable. This is not to say that the higher frequency bands are not valuable assets, but they lend themselves to a much more limited range of business plans than do the bands encompassed by the newer 802.16a standard.

As indicated elsewhere in this book, microwave transmissions above 10GHz are strictly line of sight. They have no ability to penetrate walls and are seriously impeded by foliage. These characteristics greatly limit the use of these frequencies in any kind of residential play with the exception of multitenant units (MTUs).

The short ranges of transmissions at permissible power levels in these frequency ranges and the highly focused antenna beams required to reach even those modest distance also limit the types of customer that can be accessed. You simply cannot illuminate an entire sector of a city with millimeter microwave system. You instead have to target specific buildings, preferably buildings that will yield a maximum amount of customers.

I know of no current large-scale residential deployment utilizing millimeter microwave radios. When the FCC-designated bands for local multipoint distribution services (LMDS), the first commercial deployment was in fact in a residential one-way video distribution on Long Island. It was not particularly successful, and all of the initial LMDS licensees acquiring spectrum at the end of the last decade elected to use the networks for two-way high-speed data services aimed at large businesses.

Data Services in the Millimeter Microwave Region

Two-way data is still the most compelling application for these bands, but I suggest operators try to avoid the series of mistakes made by the first generation of network operators trying to offer such services.

Most of initial license holders in the United States strove to offer the same basic type of service sold by the telco incumbents, namely, T1 leased data circuits. Because T1 is the principal competitor to millimeter microwave services, it merits some discussion beyond what was covered in the previous chapter.

Circuit-Based Services: The Wrong Model

A T1 represents an aggregation of 24 56Kbps copper phone lines and affords the user a throughput of just more than 1.5Mbps, which by today's standards is verging on medium speed rather than high speed for the business user. T1 services are highly reliable as a rule and are extremely consistent in throughput speed because actual copper is reserved for the individual user; however, bandwidth is insufficient for the highest-quality multimedia. T1s also represent choke points for the extension of Ethernet LANs since Ethernet throughputs are minimally 10Mbps rather than 1.5Mbps. Higher-burst speeds are not possible over T1 connections as is the case with many of the newer packet services, and the circuits themselves must be manually provisioned, at least on the older synchronous optical network (SONET) equipment still used by most carriers to anchor their T1 lines.

Regarding the relationship between T1s and SONET, a T1 connection occurs over twisted-pair copper. The copper pairs themselves terminate at either a central office, more commonly called a *central exchange* outside the United States, or at a digital loop carrier (DLC), an aggregation device generally placed outdoors in a hardened enclosure. The digital loop carrier itself will normally have a fiber-optic interface through which it will communicate via fiber with a central office. SONET as a standard relates to the fiber portion of the local telecommunications plant only, and SONET itself is a set of protocols and standards pertaining to the assignment of network traffic to circuits and to the restoration of the network in the event of either a failure of one of its nodes or a break in the fiber. Incidentally, SONET networks are usually set up as point-to-consecutive-point sequences of nodes that together form a closed ring. SONET supports T1 data services as well as circuit voice services, but T1s existed long before the SONET standard was promulgated.

What SONET really has going for it is redundant fiber paths and very fast restoration of service (less than 50 milliseconds) in the event of a fiber break. T1s themselves are not similarly redundant, but the fact that the core network is so reliable lends a high degree of reliability to T1 services as well. Where it is lacking is in flexibility and bandwidth efficiency.

Interestingly, the newer next-generation SONET does permit the aggregation of T1s to afford the users' higher speed, but relatively little next-generation SONET is in place in the networks as yet. Also, most installed SONET gear does not permit the rapid provisioning and activation of T1 lines. Finally, SONET does not offer inherent support for user subgroups, VPNs, transparent LANs, or other specialized service offerings.

All of this discussion of SONET is not intended as digression. It is very much to the point because the equivalent of a SONET-based T1 is what the first millimeter microwave carriers tried to provide. This was extremely misguided insofar as millimeter microwave has the bandwidth and the flexibility to offer flexible, rapidly provisionable fast packet services that form a real alternative to SONET.

The central problem facing the millimeter microwave pioneers was that T1 emulation did not play upon the real strengths of millimeter microwave and in fact resulted in an offering that was inferior to a real T1. The first generation point-to-multipoint wireless networks lacked restoration mechanisms when airlinks failed and generally offered availability levels that were at least an order of magnitude worse than those of the wireline incumbents. One equipment manufacturer, the now-defunct Triton Network, did in fact architect a SONET-like radio ring network with built-in restoration mechanisms, but the system was not widely used.

T1 emulation in my estimation is a poor strategy and one almost certain to fail, especially if it is embedded within a larger strategy of price competition, as was commonly the case among the first LMDS providers in the United States that tended to price their 1.5Mbps services at about 30 percent below the going rates for T1s in their markets. Telco incumbents can easily drop prices temporarily to fend off interlopers, and they do so. Since they are in a far better position to sustain losses to secure market share than are startups, they generally succeed.

So if offering T1 surrogates is not the answer, what is? Currently the best strategy in competing for business data customers is to offer highly flexible packet services.

Packet-Based Services: A Business with Potential

As with lower microwave, the network operator has a choice between IP and Ethernet. The 802.16 protocol is built around IP, and IP is better suited for supporting quality of service, service differentiation, and service-level agreements. On the other hand, Ethernet is the protocol used in most internal enterprise LANs today and may be delivered over low-cost interfaces. Some wireless base station equipment has the ability to support both protocols in the last mile, and for certain operators that may make the most sense. IP and Ethernet are highly compatible, and Ethernet streams are easily transported over IP networks with little overhead or wasted capacity, but the network operator is still left with the choice of whether to opt for IP or Ethernet VPNs and of what type of packet telephony standards will be supported. Thus, choosing to support one protocol to the exclusion of the other on the access level imposes certain constraints on the network operator and the subscriber.

My belief is that IP is better suited to service creation than is Ethernet but that Ethernet is better suited to service delivery—in other words, delivery of an IP stream over an Ethernet interface appears to make the most sense at this time in terms of both versatility and cost effectiveness.

Whatever packet protocol is ultimately employed to manage the link between the base station and subscriber premises, the network operator should strive to differentiate the basic access service from T1 by the following means:

- The service velocity should be far higher than is the case for T1; that is, the operator should be able to provision the service rapidly, preferably within hours of an order being placed. T1 services typically involve a lengthy provisioning process often extending over several weeks. Providers of such services are highly vulnerable to competition on that account.

- Ideally, the service should lend itself to self-provisioning in the case of changes of service. If the subscriber wants to lease additional bandwidth for a short period of time to meet special needs, then the operator should permit this through a secure Web site with automatic billing processes registering the event.

- The service should allow for the leasing of fairly fine increments of bandwidth. In a SONET network the service is generally sold in extremely coarse increments, with DS3 (45Mbps) representing the next level of service above T1. Surveys indicate that most small- and medium-business users are interested in services with throughput rates from 2 to 10 megabits per second—faster than T1 but significantly slower than DS3.

Finally, the service provider should offer the same kind of options that have been traditional in ATM networks, including committed bit rates, bandwidth reservation where appropriate, and burstable rates where requirements for high bit rates are intermittent. Guarantees regarding jitter and latency are also desirable service offerings, particularly where the network must support conferencing and multimedia or a high degree of real-time interactivity. These are areas where circuit services such as T1 excel since circuits are by their very nature highly predictable, but such benefits are, in large measure, obtainable through some of the newer IP ancillary protocols such as DiffServ, RSVP, and MPLS as well as through specific QoS provisions within the 802.16 standard. Chapter 7 covers QoS in greater depth.

Telephony in the Millimeter Microwave Bands

Compared to the lower microwave bands, the millimeter microwave region is fairly well suited to doing voice telephony as well as data. The relatively large amounts of bandwidth available means that the operator is not faced with difficult trade-offs involving data services, particularly if IP softswitches are used in lieu of circuit switches. As is the case with data services, the operator should be striving to offer what is not commonly available from the wireline incumbent. IP centrex where PBX-like functionality is offered to small- and medium-sized businesses at reasonable cost is a good example of a new type of service offering that can differentiate the competitive service provider.

Grid Computing: The New Killer App?

Grid computing is a term for a cooperative arrangement whereby a large number of collaborators make their computing resources intermittently available over a high-speed network for parallel computing applications requiring extremely high computing power. Grid computing is strongly backed by such industry giants as IBM and Sun Microsystems and has seen limited use in the financial industry and in scientific research, both for-profit organizations and academic organizations. Incumbent telecommunications carriers have expressed interest in the concept as embodied in a kind of hosted service offering where the carrier would manage user resources for a fee and ensure security, but to date no carrier has introduced a grid computing service.

In certain locales where large numbers of research institutions reside, an enterprising wireless service provider could sell a combination of high-speed point-to-point millimeter microwave links and grid computing services to a small number of high-value users. In the future, you could see grid computing making expert systems possible; a mass subscriber base could use them for various kinds of practical research and for personal reasons. You could, for example, do investment research online and take advantage of the same kind of powerful computing resources used by large firms.

All this is highly speculative at present, and I offer it only as possibility and as a stimulus to fresh thinking on the nature of service offerings in the years to come.

The ASP Model

An application service provider (ASP) is an entity offering hosted services for subscribers. These may include Web hosting and mirroring, storage, data backup and disaster recovering, security, content management, and so on. A few years ago great things were predicted for ASPs, but many failed in the marketplace, and few broadband access providers adopted the model. Perhaps a better model is to attempt to partner with companies such as IBM that offer comprehensive suites of business applications and to act either as a value-added reseller (VAR) or simply a venue for such services.

The Carrier's Carrier Model

In the carrier's carrier service model, the broadband wireless operator resells capacity to others and does not directly sell services or sign up individual subscribers. In general, this has been a low-margin business and one that has not been particularly successful at the metro level, though it is the established model in the long-distance market. In certain circumstances, ceding capacity to resellers may make business sense, but it limits the profitability that the broadband wireless operator can derive from the network.

Looking to the Future: The Importance of a Service Orientation

The ability to offer distinct services is the only sure way to differentiate yourself in the broadband marketplace, and the fact that the access providers with the greatest range of services (namely, the cable operators) have also been the most successful is no coincidence. In the future, new offerings will emerge such as location-based services associated with mobility, highly personalized content and reporting services, advanced filtering, and others that have scarcely even been envisioned. Not all new services will be successful, but those that are will serve to position the carriers that have embraced them successfully in the marketplace.

Exploring the complete range of services that broadband will support in the midterm and determining their impact is quite impossible in a book of this sort, and I can say only that tracking developments in this area is essential to the long-term health of the broadband wireless service provider. Above all, network operators must come to view themselves as marketing agents and not as public utilities. Public utilities are products of the era of regulated monopoly. In the new era of competition, they are anachronisms.

CHAPTER 4

Setting Up Physical Infrastructure

This chapter discusses the initial stages of the process of putting the network in place with an emphasis on mapping networks, selecting equipment, and siting terminals and aggregation facilities. I will leave the details of configuring individual terminals and running specific applications and services to later chapters.

Looking at the Nuts and Bolts: The Issue of Carrier-Grade Infrastructure Equipment

The information in this section is perhaps more pertinent to operators of 802.16a equipment than to those using 802.16 millimeter microwave gear, but in a sense it is relevant to all broadband operators inasmuch as it touches upon an issue that is central to the positioning of competitive access providers utilizing new networking technologies vis-à-vis the telco incumbents.

Telecommunications industry professionals like to distinguish between equipment that is known as *carrier grade* or *carrier class* and that which is *enterprise grade.* The distinction is as much about the attitude of the network operator as it is about design details of the equipment itself.

Equipment intended to be carrier grade is designed to meet the requirements of what are known as *common carriers,* a term applied to providers of traditional telecommunications services. Common carriers tend to want heavily standards-based, highly redundant, over-engineered equipment that is based on highly reliable computing platforms and utilizes closed architectures and operating systems and traditional programming languages. Such equipment must be tested and approved by industry certification bodies and must be backward compatible with legacy infrastructure. Because of the stringent performance requirements and limited size of the marketplace, carrier-grade equipment has been expensive to produce, and although its manufacturing has not been a monopoly business, it has been confined, for the most part, to a few large companies offering fairly similar products and, incidentally, not inclined to compete aggressively on price.

The requirements imposed on the manufacturers reflect the basic business philosophies of traditional telcos, which stress excellent service within a monopoly marketplace. Flexibility, modularity, low pricing, and easy programmability—or even ease of operation—have not been stressed because such attributes were not necessary to succeed in a public utility business, such as telephony throughout most of its history.

Enterprise grade, as telecommunications professionals see the issue, is a relaxed standard applying to equipment serving large private networks within corporations and government agencies. Such equipment is built to a price and sacrifices some reliability to meet a given price point. Because it is sold not to a few quasi-monopolies but to a myriad of different types of organizations, it is subject to considerable variation in design and has attracted numerous manufacturers both large and small.

Enterprise-grade equipment has been characterized by falling prices, a rapid succession of generations and standards amendments, ever-expanding feature sets and capabilities, and the use of open standards, interfaces, operating systems, and programming languages. Such equipment is increasingly self-configuring and self-provisioning.

To date, the majority of broadband wireless metropolitan public networks set up have used enterprise-grade rather than carrier-grade equipment; that is, they have tended to rely on wireless local area network (WLAN) equipment. Prior to the confirmation of the 802.16 standards, relatively little equipment that could remotely qualify for the carrier-grade designation existed, and what little existed was costly in comparison to enterprise-grade equipment, most of which was based on the older 802.11 standard for wireless Ethernet.

I have already touched upon the limitations of the 802.11 standard in respect to public services in the discussion of wireless broadband standards in Chapter 1. Inherent in the standard itself are capacity, range, and quality of service limitations that discourage using such equipment in public service networks. That such equipment is relatively less robust, redundant, and reliable than equipment designed for the carrier market also gives one pause regarding pervasive public access deployments. Finally, the fact that almost all public networks of any size attempting to operate with private LAN equipment have failed suggests that the strategy of employing it is fundamentally wrong. Perhaps the latest generation of modified 802.11 equipment might fare somewhat better in the marketplace, but this has yet to be demonstrated.

This should, it seems, put the matter to rest. But, as it happens, the issue is not so simple.

Clearly the intent of this book is to promote the new 802.16 standards and to warn against attempts to substitute wireless LAN equipment in an application for which it was never designed. At the same time, I should state that WLAN equipment can have a place in a network that is 802.16 based.

Already, 802.11 equipment is finding a prominent place in a certain rather specialized type of public network familiarly known as a *hotspot*. Hotspots may be considered to be the Internet equivalent of a pay phone—a short-range access point situated in a public space that would enable an individual with an 802.11-capable phone or computing device to access the Internet after an online credit transaction. Some tens of thousands of these hotspots are already scattered across the globe, and their number is increasing daily. Within the overall category of broadband public networks, they represent the greatest success story thus far.

The installation of hotspots represents a legitimate strategy for the broadband wireless metro operator, though it should be considered as merely one service offering. Network operators may also choose to provide backhaul services to other hotspot operators, and in this market they may be able to compete effectively with the digital subscriber line (DSL) and T1 services favored by most hotspot operators today.

Also, 802.11 airlinks may be useful for in-building extensions of the public network as in multitenant units (MTUs) and business parks, and their installation may be offered as a value-added service by 802.16 network operators.

What I do not advise is an attempt to deploy 802.11 WLAN equipment as the core infrastructure in a wireless broadband metropolitan network. Such a move is a false economy and is almost certain to result in a degradation in service that will ultimately cause the network to fail as a commercial entity. As indicated previously, a few products in the marketplace are based in part on the 802.11 standard but are provided with ancillary protocols and specialized interfaces that do represent legitimate attempts to modify the technology for use in the metro; however, such products almost invariably cost substantially higher than WLAN equipment, often double or triple the cost. Therefore, a decision to utilize equipment designed from the ground up for use in the metro rather than modified office equipment appears to make more sense.

Obtaining Roof Rights, Right of Way, and Access to Appropriate Buildings at Acceptable Cost

Almost every major wireless broadband operator faces the same challenge when planning and building the network. How does one manage to place network nodes where they are needed to serve the subscribers?

If one does not meet this challenge early in the deployment process, one's chances of succeeding with the operation will be nil. In the business of broadband services, access is everything—the absolute precondition to moving forward with provisioning and ultimately toward profitability.

Access requirements for the network operator will vary somewhat according to the frequencies at which the network is operating, the types of customers sought, and the capabilities of the equipment. A network targeting primarily MTUs and business high-rises will need to get roof rights for the individual network nodes as well as the central base station(s). A network that is serving residential customers will not. A network using strictly line-of-sight equipment, will, all things being equal, require a denser infrastructure of base stations and thus will require more roof rights or tower rights than a system using the newer non-line-of-sight (NLOS) equipment. The cell size of the NLOS network will be smaller (see Chapter 5), but, on the other hand, each base station will be able to reach more potential customers located within its effective radius. To date, networks using line-of-sight equipment and operating in developed markets have nearly always failed because the expense of building and maintaining a large number of base stations has proven prohibitive.

Central Office and Main Base Station Facilities

Obviously, the first order of business is to establish a site for the initial base station and the central office for the entire network operation. The central office need not include all the administrative facilities of the network, but it should include the equipment essential for anchoring the network, which would ultimately depend on the mix of services offered by the network.

Several considerations are significant here.

The antenna for the central base station needs to be situated at a position and at an elevation where it will encounter the fewest obstructions between itself and targeted subscriber sites and, as the network is built out, subsidiary base stations. This is true even if NLOS equipment is employed, because a clear line of sight is always preferable.

Finding the right location can be difficult, though. A building that is considerably taller than any of its neighbors or a peak overlooking the city or metropolitan areas is often ideal from the perspective of minimizing obstructions, but it may be far from ideal in other respects. Space within the city's tallest structure may simply be too expensive to rent, and a mountaintop is likely to be too remote to permit the placement of central office facilities. Another option is to place a steel antenna tower atop a shorter structure, but often local ordinances will not permit this. One can, of course, run a high-speed connection to a remote antenna, the usual practice among radio and television broadcasters, but that connection will have to be capable of handling all the traffic going through the hub with no bottleneck or reduction in throughput speed. A fiber connection of that type may be unobtainable or too expensive, and a wireless connection will have to be capable of fiberlike speed and of maintaining a high level of availability. This may entail leasing millimeter microwave backhaul from another service provider if one is operating in the lower microwave region.

To obtain affordable space for central office facilities, one may have to settle for a location that is less than optimal in respect to signal distribution and customer acquisition, and such considerations will inevitably have an impact upon the network's growth potential. Because the choice of central office and initial base station placement is so important, it is best for the network operator to select a site prior to proceeding any further with the network and make certain that it can be secured under favorable circumstances before constructing a single access point.

Physical Requirements for the Central Office

During the acquisition process, careful attention must be given to space, power, and ventilation requirements for the basic central office equipment as well as to the positioning of the antenna. The operator will need to know how many standard rack units of space will be taken up by the essential equipment and should have a fairly firm notion of what types if not brands of components will be purchased prior to network launch. Chapters 4–6 provide guidelines on what to look for in terms of equipment specifications.

At the least, the operator will require the following components for delivering even the most basic service offerings.

First, operators need a base station transceiver attached to the antenna. In some cases, this may be placed in a hardened enclosure adjacent to the antenna and in some cases in an equipment room or closet. Since cable that runs from the radio to the antenna should be kept as short as practical to avoid signal losses, it is not advisable to place the antenna on the roof of a 50-story high-rise and the transceiver in the basement. Typically, such transceivers are considerably larger and more powerful than subscriber units, and they feature redundant architectures and backup power. Therefore, they require a certain amount of space, preferably a sizable fraction of a cubic meter.

Most broadband wireless systems also include a special component that incorporates essential software for managing the physical layer and may also include routing or switching software. In all cases, this box will include what is known as *wireless middleware*, which serves to modify the Transmission Control Protocol/Internet Protocol (TCP/IP) data stream so as to adapt it to the airlink. The design of such boxes is far from uniform, and in the first generation of lower microwave equipment, a modified Data over Cable Service Interface Specification (DOCSIS) cable data headend device was often employed. Today that approach is thoroughly discredited, however, and indeed with the finalization of 802.16 and the production of equipment embodying it, the network operator need not even consider such a makeshift. Generally, the base station transceiver will be combined with the router/switch.

Second, the operator will also need an edge router whose output will go directly to the radio transceiver, if the two are not combined in a single box. The size and power consumption of this device will depend on the number of subscribers served. Most edge routers made today are highly modular, consisting of a largish box with multiple slots taking a number of separate blades, each of which handles a group of input/output (I/O) ports. The router is designed for a certain maximum number of ports, and in most cases the operator chooses to buy additional blades as needed until reaching the full capacity of the design. A fully loaded edge router will, as a rule, take up several rack spaces and will have a power consumption in the kilowatts. Carrier-grade edge routers cost thousands of dollars, but a small metropolitan network may be able to get away with an enterprise-class router, which costs less than $1,000. In some cases, as I have mentioned, the base transceiver will incorporate an edge router. Figure 4-1 shows a Juniper edge router.

Figure 4-1. *A Juniper edge router, courtesy of Juniper Networks*

Some products that are primarily edge routers perform other functions as well, such as creating services, inputting and outputting protocols other than IP, and providing switching functions within those other transport protocols. Such "Swiss Army knife" network elements are popularly known as *godboxes* and are intended to provide the network operator with lots of choices and a high degree of flexibility.

The utility of such devices is a matter of some contention in the telecommunications world today. Little uniformity exists in the design approaches embodied in godboxes, and any network operator contemplating using such a device must carefully study its capabilities. Depending on the mix of services the operator intends to offer, some such devices may be useful. But typically the versatility comes at a price. The boxes incorporate custom designs and often utilize proprietary elements, architectures, and engineering approaches and, in many cases, represent staggering development costs that cannot be amortized even among a large universe of users. Furthermore, none of the godboxes on the market today have been designed with the specific needs and requirements of the wireless operator in mind; instead most are designed to interface with optical networks.

Another essential element in the central office equipment rack is a server devoted to the subscriber database. The same element may, depending on the size of the network and the desire of the operator to assign the various networking functions to discrete physical platforms, also contain the billing and provisioning software and may host the authentication software as well; however, more commonly, in the interest of ensuring the highest degree of security, authentication will be performed on a separate server. Radius authentication software has become the de facto industry standard for telecommunications and large enterprises.

The central office may also, again depending on the mix of services, contain such as elements as the following:

- A softswitch
- An IP telephony gateway (occasionally the two elements will be combined on one physical platform)
- A content server for supporting "walled-garden" applications that are not resident upon the public Internet but are available only to those who subscribe to the wireless broadband network
- A video server for caching multimedia material to be streamed to subscribers or else accessed on demand, or, alternately, for ad insertion
- A server for content management software
- A hardware security device for bulk encryption/decryption
- A satellite transceiver for accessing content
- An optical transceiver for interfacing with a metro ring or mesh
- A DSL access multiplexer (DSLAM) for interfacing with DSL links

Certainly other network elements are possible as well, and conceivably the central office for a large metropolitan wireless operation could have two or more floor-to-ceiling racks filled with equipment.

Any central office, large or small, should occupy a secure location where access to the facilities is strictly controlled and the facilities are monitored at all times. In today's political climate considerations of physical security are not secondary, and the days when major Internet access points were left untended in unlocked rooms in parking structures are long gone.

The central office should have an uninterruptible power supply (UPS) or supplies capable of powering the central office equipment complement for at least 48 hours. Such backup power may utilize banks of batteries, fuel cells, internal combustion generators, or combinations thereof. The important point is that the backup power supplies deliver dependable AC power at a fixed voltage and with low values of line harmonics, preferably less than 1 percent under conditions of load. In this context, the low-cost UPSs utilized in business offices are unacceptable. Medical-grade backup power should be the standard, with appropriate power conditioning apparatus to maintain a clean, smooth 50- or 60-cycle sine wave, and the system should be designed to provide *ride through*, that is, continued delivery of electrical power while the backup power-generating system is coming up. The specifics of backup power facilities design are quite involved and are beyond the scope of this book, but the objective is simple—to ensure that the central office will continue to operate perfectly in the case of a power blackout or brownout. My recommendation is that the network operator retain a consulting engineer with demonstrated expertise in power quality.

The facility where the vital equipment resides should have personnel on the premises at all times to control access and should be provided with ancillary surveillance and alarm systems communicating back to a highly secure monitoring facility.

Equipment should be professionally mounted on steel or aluminum racks, and cable management accessories should be employed so that reconfigurations and additions to the network can be easily managed. All equipment should be easily accessible to technicians, and racks should be situated with sufficient clearance to permit any connection to be manipulated without the removal of a component from its rack. A centralized command console with a high-definition monitor and comfortable seating for the network manager is advisable.

The equipment racks should rest on a raised floor as a safeguard against natural disasters. The interior should be climate controlled and properly ventilated, and auxiliary independently powered climate-control systems should be in place in the event of a power loss. The structure enclosing the equipment should be highly fire resistant, and highly impact-resistant glazing should be installed. Doors and locks must be completely resistant to being forced open with hand tools.

All operating software and customer records must be backed up continually and mirrored at a secure facility. Retaining a reputable disaster recovery firm to mirror the entire computing operation at the central office is also recommended in the event of a large-scale catastrophe that destroys all or part of the central office.

All these recommendations constitute best practices and are entirely typical of traditional telecommunications incumbents, but they are not observed by all broadband access providers, and the degree to which they are absolutely essential is debatable. If one is providing nothing more than high-speed access to residential customers, then one may reasonably opt for a lower degree of security and infrastructure integrity at the central office, though one would certainly have strong reason to protect customer records. If, on the other hand, one is providing essential telephone service, then one incurs a grave obligation to the subscriber and to the larger community. Business users, particularly larger enterprises, will also expect that the network will be well secured and equipped to survive disasters. The loss of a data link for hours or days can be devastating to a business, especially one involved in conducting online financial transactions and recording such transactions. The degree of legal liability facing a carrier or service provider that fails to follow such best practices is uncertain, but no sane individual would want to compromise the business of a major subscriber or jeopardize public safety because the network failed to perform adequately.

Obtaining Central Office Facilities

Network operators can obtain suitable central office facilities in a number of ways. They may simply utilize leased space in a commercial building after obtaining the owner's permission to make the necessary physical modifications to house the equipment. They may purchase a building or an office condominium. They may lease rack space in a large data center or "telco hotel." Or they may collocate in the central office of an incumbent carrier.

The first two ploys entail heavy initial expenditures for construction and installation. If the central office is of any size, and the operation lacks staff with the requisite installation experience, the network operator may need to retain the services of engineering design and construction firms specializing in building network hubs, and such services do not come cheaply. Design and installation fees will generally run well into the tens of thousands of dollars, but I cannot be much more specific because installation requirements vary so markedly from one site to the next. Such services can and should be solicited through competitive bids, but price should be only one consideration. Special expertise is required for telecommunications installations, and firms whose experience has been limited to small corporate LANs are likely to be deficient in skills and experience.

Commercial data centers and telco hotels can save the network operator much time and labor because many of the necessary amenities are already in place, particularly in the case of the latter. Physical security, backup power, and the situation of the equipment racks will have already been addressed in most such facilities, and varying connections to the public switched telephone network (PSTN) and the Internet will already be in place. The more elaborate facilities may even broker specific arrangements with owners of long-haul fiber for ensuring that service-level agreements can be maintained and that latency through the Internet can be confined to a certain maximum value.

Leased facilities of this nature have little uniformity. A telco hotel, for instance, is an operation that has been designed from the ground up to serve the needs of service providers, particularly competitive service providers. Other ventures specialize in serving Internet service providers (ISPs) or in providing Web hosting facilities for private companies, and they are not apt to stress long-distance peering to the same extent. Still others are intended to provide space for enterprises to place their data centers or data storage equipment, and these may provide no carrier-to-carrier interconnectivity.

The problem with all such entities is, of course, recurrent cost. Well-constructed and well-managed data and telecommunications centers are expensive to build and operate and often occupy prime real estate, thus imposing high recurrent costs on the operators of these centers—costs that must be recouped by charging tenants top dollar. And, at least in the United States, the number of such centers has been diminishing sharply after reaching a peak in 2001. With the wave of bankruptcies afflicting the dot coms and competitive local exchange (CLECs), both of which had acute if often temporary needs for such facilities, the business case for operating them grew steadily less favorable. Still, a number survive, and for the wireless network operator attempting to launch a network, they may fulfill a need. I suggest, however, that the ownership of one's own central facilities is a desirable long-term goal.

As for the collocation of one's central office facilities in someone else's central office, avoid it. Telco incumbents do not want to have interlopers on their premises, and if forced to do so by regulation, may do all in their power, up to and including outright sabotage, to impede the competitor, and, as you have seen, DSL independents have made numerous allegations to that effect. In any case, a wireless operator has far less reason than a DSL operator to be in a telco central office and should not contemplate doing so in spite of the excellent facilities maintained by the incumbents.

Additional Base Stations

Once network operators have secured central office facilities, they must then turn their attention to the matter of access points. Initially, the network may have but a single access point that will be situated within or adjacent to the central office, but, as the network expands, additional access points will be required. Determining precise requirements in this regard is no simple matter, but it is essential to the success of the network.

Several factors should play a role in the optimal siting of access points in the network, including the nature of the equipment (line-of-sight or NLOS and beam-forming antenna, or lack thereof), the effective range of the access point transmitter, the number of potential customers within range of the transmitter, the extent to which frequency reuse can be achieved within a given area, and the availability of suitable locations at an affordable price. None of these factors should be considered in isolation.

A network based on a large number of NLOS subscribers is going to have to put the subscribers closer to the base stations—in other words, more base stations will be needed. Density of the subscriber population and the reach of the transceivers will also have a bearing on where base stations are situated. The first Multichannel Multipoint Distribution Service (MMDS) networks that transmitted over licensed spectrum at relatively high power and could maintain links over distances exceeding ten miles could serve thinly scattered customers over a large area with a single base station. Network operators using unlicensed spectrum and transmitting at 1 watt maximum power are unlikely to be able to duplicate that and instead will require a number of base stations for the same area and the same customer density. I want to generalize in this regard and say that at such-and-such customer density over so many square miles at 5.8 gigahertz (GHz) this many base stations will be required, but it is never that simple, because the same equipment may perform differently in different radio frequency (RF) environments.

Wireless network operators add base stations reluctantly—doing so only when the reach and/or capacity of the existing base stations is insufficient to serve what is seen as a sizable number of potential customers. One knows when one is approaching that point because of increasing network congestion, and one then sets out to find a location of the new base station. Insofar as possible, the network operator should attempt to plot traffic patterns at least two years into the future, however, and should plan the location of future base stations long before they are actually required.

This brings you to the final topic regarding base stations: identifying suitable locations for base stations.

Particularly in the case of line-of-sight equipment, one wants a base station with a minimum of obstructed pathways to potential customers. In most cases, the base station antenna should occupy an elevated position where sight lines extend well over the tops of the tallest intervening structures.

In recent years the tendency in American cities has been to place antennas for wireless networks on towers specially constructed for the purpose and owned by companies such as American Tower and SBA, which have developed businesses based on the lease of space to network operators. The rise of such companies has been the direct result of the proliferation of mobile telephone cell sites and the determination on the part of municipal governments to limit their number. By obliging commercial operators to collocate on few approved sites, that proliferation has been checked, and, coincidentally, a new industry has emerged.

Broadband wireless operators can certainly elect to occupy such a site, but they may not necessarily be welcome or able to afford the lease if they are. Tower owners do not charge uniform rates, but a rough average is $1,000 a month. This is nothing to a cellular operator with tens of thousands of customers, but for a struggling wireless broadband operator it is a significant expenditure especially when that operator is expected to sign a 30-year lease in the bargain. Another issue with towers is that they are primarily intended for hosting lower-frequency mobile services rather than networks operating above 2GHz, and they may not provide optimal elevation since the mobile services lack line-of-sight requirements.

If the broadband wireless network operators are disinclined or unable to gain access to a tower, they can always attempt to negotiate roof rights on a multistory building. Here it is difficult to generalize, because rates can vary tremendously. In some instances, if the network operator is willing to provide data services to the tenants of the structure in question, recurrent fees may be reduced or even waived.

Backhaul

Backhaul refers to the connection from an access point or base station back to a central office facility. In a broadband wireless network, backhaul, ironically enough, generally occurs over wireline connections, although wireless links are perfectly possible as well.

Obtaining backhaul connections at reasonable rates is essential if the network is to operate profitably, particularly as the network expands and more and more backhaul is required. For this reason the operator must determine the means of obtaining backhaul for given locations before a single network element is put in place. One simply cannot build first and then start casting about for backhaul solutions.

Backhaul itself is a rather complex subject. Several distinct physical media have been used for providing it, and a number of different arrangements are possible with companies offering backhaul capacity. One needs to explore all options during the planning stage in order to make certain that the essential backhaul component is being obtained in the most cost-effective manner.

Backhaul itself represents the aggregation of network traffic, the sum of each transmission to or from a base station and an individual subscriber node. For this reason the capacity of the backhaul must be significantly greater than the capacities of the individual access airlinks. Nevertheless, the capacity need not and should not equal that which would be required if all subscribers in a cell were transmitting simultaneously, because such an eventuality is highly unlikely. The capacity of the backhaul can be as little as a tenth of the aggregate capacity of all individual airlinks, though a four-to-one or six-to-one ratio is more prudent. If, for instance, 100 customers are each provided with a 10 megabits per second (Mbps) connection within a single cell, then the combined capacity of all of the individual airlinks is 1 gigabits per second (Gbps). A 100Mbps backhaul could suffice in that situation, though 250Mbps would be ideal.

Quasi-Backhaul

It should be noted here that there are a couple of ways to set up wireless broadband networks requiring little or no backhaul. These involve two variant network architectures, the mesh network and the ring or point-to-consecutive-point network, both of which will be discussed at further length later in this chapter.

In a mesh, no real distinction exists between a base station and subscriber premises terminal. Instead of each subscriber terminal transmitting back to a base station, which itself would define a cell, mesh terminals transmit to one another, passing the signal along until it arrives at its final destination at the central office. Each terminal will normally be endowed with some intelligence and will include a small router and/or switch for determining which adjacent node in the network will be selected as the intermediate destination for a message. The terminal will be able to choose among all the other subscriber terminals within reach and will normally compute a path that is least congested. If one of the terminals is malfunctioning, the transmitting terminal will simply route around it. A wireless mesh is a form of peer-to-peer network, but the peering takes place on the physical layer as well as layers two and three of the network stack. In effect, the mesh provides its own backhaul.

To paraphrase George Orwell, in a mesh all nodes are equal, but one node is more equal than the others, and that is known as the *seed.* The seed is the first node installed and communicates directly with the central office equipment and through it with the PSTN and the public Internet. In networks where subscribers are few and scattered, more than one seed may be required.

The Internet itself takes the form of a logical mesh and stands as the prime exemplar of the robustness and essential soundness of the mesh concept; reflecting on the explosive growth of the Internet, some equipment developers familiar with its history have assumed a certain inevitability to the mesh concept in the wireless domain as well. And perhaps such inevitability is real. Still, mesh equipment to date has achieved little success in the marketplace, and most of the more than 20 manufacturers that have developed it have been forced to shut their doors.

I suspect that price has been the major issue. Putting a router or switch at every terminal cannot but increase the cost the subscriber premises package, though it should make for a less-expensive central office installation. Against that price must be weighed the price of backhaul, which of course is highly variable. If backhaul providers are determined to gouge the wireless broadband operator, then opting for a mesh configuration is a counter move. But one should remember that there are other ways to implement wireless backhaul where the wireless broadband operator still retains ownership of the backhaul channel, so the mesh must not be regarded as the sole solution.

Ring architectures, the second means of eliminating the backhaul, are equivalent to daisy chains or buses. Each node connects to two other nodes, and the whole assemblage forms a closed ring. Rings tend to be impractical when a large number of subscriber nodes are involved and are much less robust than meshes; on the other hand, they require less network intelligence. Nevertheless, they are similarly expensive to implement because of the presence of routers, switches, or add-drop multiplexers in the subscriber terminals. Much of the point-to-point microwave equipment designed for mobile backhaul supports point-to-consecutive-point architectures as well, and one should opt for this approach when purchasing equipment.

Fiber-Optic Backhaul: The Gold Standard

If the price is right, fiber backhaul is preferable to all other methods. Fiber links offer higher throughput speeds than any competing access technology and extremely high availability. All fiber links are not equal, however, and, moreover, the fiber link should always be considered in the context of overall metro core network, of which the fiber itself comprises the lowest layer.

Fiber-optic networks in the broadest sense may be divided into two categories: active and passive. *Active* networks employ optoelectronic elements that on-load and off-load traffic and direct it along the network paths, as well as amplifying and conditioning the signals to extend distance and reduce error rates. *Passive* networks are much simpler and employ passive optical splitters to direct the signal along parallel optical paths, generally not exceeding 32 because each split halves optical power in the resulting pathways.

PONS

Core optical networks in metropolitan settings are almost exclusively active today. Passive optical networks (PONs) are mainly used as an access technology, though PON manufacturers have been trying to promote them for backhaul for years. PONs carry much lower equipment costs than traditional active optical networks, and they offer aggregate throughput speeds in the low gigabits per second with the possibility of much higher speeds through a technique known as *wave division multiplexing.* For a wireless broadband operator able to obtain "dark fiber," explained in detail shortly, a PON provisioned by the broadband operator may constitute an attractive option for backhaul. As is, the number of firms offering PON backhaul anywhere in the world probably does not exceed 20. Figure 4-2 shows a passive optical network.

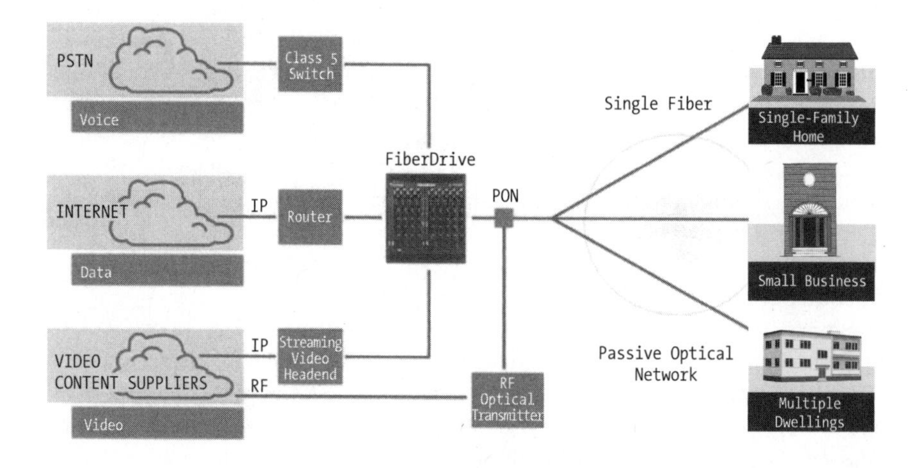

Figure 4-2. *A passive optical network, courtesy of Optical Solutions*

Active Fiber Connections

Active optical networks are based largely on synchronous optical network (SONET) and synchronous digital hierarchy (SDH), which means that available bandwidth is apportioned in temporal channels taking the form of fixed time slots. Such networks are inefficient in their use of network resources, and the services running over them are generally expensive, but reliability is high. Packet services, which are just beginning to challenge circuit in the marketplace, are generally less reliable but less expensive per a given throughput speed.

Circuit Fiber Connections: SONET/SDH: Circuit time intervals are all multiples of the basic 56 kilobit per second (Kbps) voice channel. The chief service offering within SONET networks is the T1 at approximately 1.5Mbps, and the equivalent SDH offering is the E1 at approximately 2Mbps. As indicated earlier, multiple T1/E1 services may be supported by some of the newer SONET equipment platforms, but most carriers have not installed such equipment and will not provide fine increments of bandwidth. In the United States the next level up from the T1 is the DS3 at 45Mbps, and the next increment above that is the OC-3 at 155Mbps. Someone desiring a 20Mbps or 100Mbps backhaul is simply out of luck.

Of course, one can always choose to lease more capacity than is absolutely needed, but the pricing of DS3s and OC-3s is steep—thousands of dollars a month for the former and well into the five figures for the latter. For a startup wireless broadband network attempting to build a business, such large-capacity circuit backhaul connections are a heavy burden, a burden made heavier by the fact that they are quite inefficient for the transport of packet traffic and carry a great deal of network overhead.

Packet Fiber: Metro Ethernet or IP: Another option for backhaul is a direct packet connection using either the Ethernet or IP protocol. Fast packet services over fiber have been offered by a number of startups such as Cogent and Yipes in the United States and more recently by some of the incumbents. Quite a number of carriers are now offering such services in East Asia as well, particularly in mainland China, Hong Kong, and Taiwan. Certainly in the case of the startups, the pricing has been far more attractive than is the case for circuit services with monthly fees as low as $1,000 for a 100Mbps burstable connection, but in most cases such services are not delivered with rigid service-level agreements in respect to jitter, latency, minimum throughput, bandwidth reservation, and so forth. Direct IP services better support quality of service than do pure Ethernet services, which is something to consider when evaluating fast packet backhaul providers.

Where available, packet services over fiber are often preferred to circuit connections strictly on the basis of price, but the problem is that they are not too generally available, and in nearly all cases they are offered by entities that would prefer to have the customers whose transmissions are being backhauled. Then, too, fiber-based access services are expensive to provision, and the provider of metro Ethernet or direct IP-based services may simply not be able to lay fiber to the building being served by the broadband wireless operator.

Dark Fiber: Another option, sometimes present and often not, is the lease of so-called dark fiber from a public utility or public transit system. Gas and electrical utilities, and, to a lesser extent, transit districts and railroads, own extensive amounts of optical fiber that they use for monitoring purposes and internal communications. Such private fiber networks are generally grossly underutilized—largely *dark* in the parlance of the telecommunications industry—and the owners are often amenable to leasing capacity, in which case the lessee may be assigned an individual wavelength or an IP address. Since utilities are not primarily in the business of selling fiber capacity and do not regard such networks as major profit centers, they can often be persuaded to lease capacity at reasonable rates. Be forewarned, however, that using leased dark fiber for backhaul or for any other purpose is a different proposition than purchasing a T1 or fast packet service from an optical services provider. Public utilities often do nothing other than provide access to an optical pipe. They do not provide termination or optical conversion equipment, and they offer no guarantees in respect to network redundancy or provisions for restoration. Consequently, by accepting such an arrangement, broadband wireless operators are in effect becoming optical carriers as well as wireless access providers since the burden of operating and maintaining the optical portion of the network will fall squarely on their shoulders.

Stranded Fiber Assets: Yet another possibility is securing what are known as *stranded fiber assets.* During the late 1990s and the beginning of this decade, a tremendous amount of fiber was laid, particularly in the United States, on both the local level and between cities. Such overcapacity led to falling prices for leased fiber and a wave of bankruptcies among owners of fiber infrastructure. Quite a bit of fiber laid by bankrupt companies is now available for resale from fiber brokers such as Fiberloops, often at extremely low prices.

Such stranded assets can be an enormous bargain for the startup broadband access provider, but their usefulness is subject to several qualifications. Usually the optical fiber that was installed did not form a comprehensive or pervasive network but more commonly took the form of a core ring that served to anchor what are known as *laterals,* fiber runs to individual buildings. Laterals, the access portion of a fiber network, are by far the sparsest fiber deployments and reach less than 10 percent of business locations in the United States. From any given base station a fiber lateral may simply be unavailable.

One can, of course, situate one's base stations with a view to exploiting extant fiber resources, if one can determine their whereabouts. A number of consultancies such as the aforementioned Fiberloops and TeleGeography offer databases of fiber resources, but no one pretends that any such database is exhaustive. Because fiber is an inherently valuable resource in and of itself, requires no licensing, and has proven itself to be long lasting and reliable, and thus far has supported steady increases in throughput speeds through upgrades and improvements in the terminal devices, acquiring fiber is generally a wise decision. But unless the fiber extends the entire distance between the points to be linked, it is fairly useless for backhaul.

The network operator who can obtain only some portion of the fiber required to complete a backhaul connection may choose to construct the remaining portion required. This is not an operation to be undertaken lightly and in most cases will be prohibitively expensive. Trenching costs for fiber builds can run as high as several hundred thousand dollars a mile in a large metropolitan area and several tens of thousands in more rural areas. In some cases fiber may be hung from utility poles or even snaked through sewer or gas lines, both installation methods being roughly half the cost of trenching. Still, installation in the best of circumstances is seldom cheap and, moreover, usually entails a long and involved permitting process. Fiber builds are also time consuming as a rule.

Other Physical Media for Backhaul

While fiber is certainly the preferred medium for backhaul in respect to speed and availability, it is far from the only option, and frequently it is far from the most cost effective.

Free-Space Optics for Backhaul

A further possibility is obtaining a fraction of the fiber needed on a lease or ownership basis and utilizing free-space optics to provide fill-in. One manufacturer, LightPointe, makes a free-space optical system where the transceiver can link fiber sections transparently with no need for optical-electronic-optical (OEO) conversion, always an expensive proposition.

Hybrid Fiber Coax

In a few locations, mostly in the United States, a new type of hybrid fiber-cable system operator known as an *overbuilder* may offer packet services capable of speeds in the 10–30Mbps range, sufficient for backhaul in some instances. Few such operators guarantee quality of service, however, and here again the cable operator is likely to be competing for the same access customer as the broadband wireless operator.

VDSL

VDSL, with speeds in excess of 50Mbps in some cases, might also be considered for backhaul where available, but availability is limited as yet. Even more limited opportunities exist for utilizing a power line carrier for backhaul, a technology where data is transmitted over AC power lines at throughput speeds as high as 30Mbps. One New England–based company named Amperion is pursuing a strategy of combining powerline carrier backhaul with wireless broadband access, but to date it has achieved few commercial deployments. Bear in mind that any such arrangement requires the active participation of the local electrical utility, and while such entities have long evinced interest in gaining a stake in the communications business, few have made major commitments to doing so.

Wireless Bridge Connections

Wireless broadband operators, can, if they so choose, pursue a wireless pure play in regard to backhaul and utilize wireless point-to-point "bridge" connections from the base station to a central office. They can do so using their own spectrum, or they may elect to use another wireless service provider, preferably one utilizing millimeter microwave equipment or free-space optics.

Such bridge connections can utilize either low microwave or millimeter wave frequencies and will normally employ the full available spectrum in the one airlink. Very high-gain, narrow-beam antennas are the rule here, and maximum transmission distances are multiples of the radii of the cells being backhauled. Wireless bridge backhaul connections exceeding 30 miles are feasible in some instances.

The same spectrum can be used for backhaul as is used for access, but the prevailing practice is to use a dedicated band to avoid interference since the backhaul has the potential to interfere with every subscriber in the cell inasmuch as it reuses all of the spectrum. The 5.8 U-NII band is frequently assigned to backhaul and offers a good combination of bandwidth and range—100 megahertz (MHz) and more than 20 miles, best case. Other frequencies favored for backhaul purposes include the 18GHz and 23GHz licensed bands where licenses are fairly easily obtainable in the United States, the 24GHz unlicensed band, the 28GHz–31GHz Local Multipoint Distribution Service (LMDS) bands, the 38GHz and 39GHz bands, and the unlicensed band at 60GHz. In all of these bands at least 100MHz of spectrum is available and interference is minimal. Range is an issue, however, especially as one goes higher in frequency, and in the highest band, centered at 60GHz, distance should not exceed a kilometer.

The cost of radios for the millimeter wave regions is significantly higher than for the lower microwave bands, and bridge links are minimally several thousand dollars apiece. In the case of LMDS, 38GHz, and 39GHz, the cost is apt to be much higher, approaching $100,000, though that must be balanced against the generous allocations of bandwidth and resultant high throughputs and the fact that the bridge equipment represents a one-time capital cost along with the generally fairly minimal recurrent costs associated with roof rights.

THE NETHERCOMM SOLUTION

A unique wireless broadband technology has been recently introduced by a California company called Nethercomm. At this point its possible impact on the market cannot be determined, but the technology forms an interesting option that may provide real benefits in certain networks.

Nethercomm's system is based on ultrawideband (UWB) radio technology where the whole available radio spectrum from the low kilohertz to the tens of gigahertz is utilized. Unlike more conventional over-the-air UWB systems, Nethercomm's radios transmit over *waveguides*, closed metal pipes that contain the signal and prevent interference with other radios. The idea of using waveguides for propagating radio transmissions is not new, and beginning in 1947 much of the long-distance telephone traffic in the United States took place over underground waveguides. What's different about the Nethercomm system is the use of UWB in lieu of conventional modulation techniques and the notion of employing infrastructure that is already in place, namely gas lines.

Nethercomm claims that their prototype systems are capable of outputs of several gigabits per second, faster than most passive optical systems, and that the outlay for equipment is far less. If these claims are true, the system could succeed, though it is arriving in the marketplace critically after many rival technologies have already established themselves.

Of course, such a network depends on forging an agreement with the local natural gas utility, a process fraught with uncertainties. As is, most gas utilities with an interest in broadband have opted to exploit dark fiber or install new types of armored fiber inside the gas lines.

Free-Space Optics Pure Plays and Hybrid Networks for Backhaul

The final option for wireless backhaul is free-space optical, briefly alluded to previously in the text.

Free-space or free-air optical is a technology for transmitting information by means of infrared laser pulses without the medium of optical fiber. Speeds equivalent to those of fiber are attainable in theory, though existing systems are not much faster than millimeter microwave. Free-space optical equipment does not require a license to operate and, while generally more expensive than low frequency microwave, has been coming down in price. A single point-to-point link can run as little as a few thousand dollars or as much as several tens of thousands of dollars.

Free-space optical equipment can be tricky to set up and maintain because transceivers must be precisely aligned with one another, and something as minor as the settling of a building can severely misalign a link. Autoaligning equipment does exist, but it tends to be expensive. Condensation on the lens, dust, and soot can also be problems. As with other wireless transceivers, siting is critical, and roof rights are generally required. Some systems can be operated through window glass, though distance is critically reduced in such instances, and metalized window coatings effectively prevent such transmissions altogether.

Free-space optical systems can be used in tandem with millimeter microwave—the combination of 60GHz and free-space is especially intriguing because the two are well matched in speed and have complementary attenuation characteristics. Obviously such a ploy is expensive, but it provides the highest speed and best availability of any wireless link.

T1 Backhaul over Copper

I mention this unsatisfactory expedient, T1 backhaul over copper, only because it has been used in the past. A T1 connection affords the user a throughput rate of only 1.5Mbps per second, a tiny fraction of the speeds supported by 802.16. Aggregations of T1s, which are offered by many telcos, are of course better, but they are still woefully inadequate. I simply do not see how a network operator can possibly succeed with such a legacy solution and strongly advise operators to exclude T1 categorically as a method of backhaul.

Determining Basic Network Architecture

I have already mentioned the basic network architectures: point-to-point, point-to-consecutive-point, point-to-multipoint, and mesh. While hybrids consisting of one or more topologies are certainly possible, generally a single architecture is chosen for the entire locale to be served, and the choice reflects not just technical considerations but the nature and composition of the local market.

The band or bands in which the network operator is transmitting limits the choice of network architecture. For example, transmissions in the highest band now commonly in use, 59GHz to 64GHz, can never be point-to-multipoint because spreading the beam to accommodate such an architecture would result in unacceptably short range and also a failure to reach many potential customers because of strict line-of-sight considerations. Similarly, a mesh architecture, while technically feasible in the millimeter wave frequencies, is apt to be quite expensive because each node essentially becomes a base station (one U.K. manufacturer, Radiant Networks, now out of business, did make a 28GHz mesh product, however). Finally, point-to-point connections are rarely encountered below 3GHz though they certainly are possible, because the bandwidth lends itself to point-to-multipoint deployments and the profit potential of the latter is generally much greater than that associated with the sale of the total bandwidth to a single customer via a narrow-beam bridge link.

Point-to-Multipoint

For networks utilizing low microwave frequencies, a point-to-multipoint architecture will represent the norm. Point-to-multipoint will enable the network operator to reach the greatest number of subscribers at the lowest cost and will sharply limit the number of routers and switches required for the network. Figure 4-3 shows a point-to-multipoint network.

Point-to-multipoint deployments have frequently been advocated for millimeter wave frequencies as well, but few have actually been built. The problem lies in the topography of most large cities, which would be the prime markets for such services. Given the high cost of base stations, minimally $100,000 for equipment alone, operators will probably be unable to populate the market with more than two or three such facilities. But at the same time it will be hard to reach all potential customers within the radius of a given base station because of blockage. Because of the unhappy experiences of first-generation LMDS network operators in attempting to sign a sufficient number of customers to achieve profitability, many authorities today have concluded that point-to-multipoint architectures are seldom if ever advisable for millimeter wave networks.

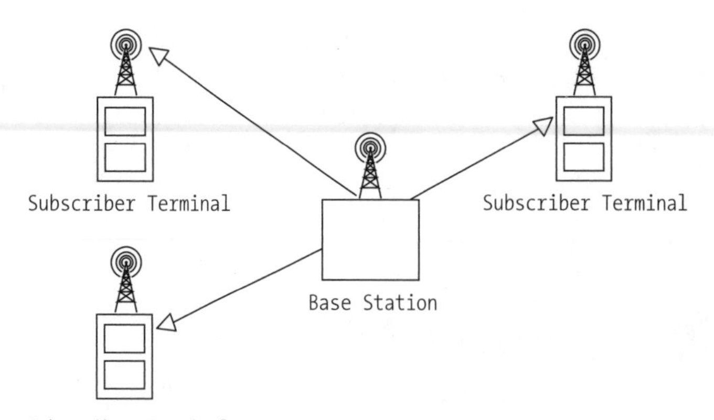

Figure 4-3. *A point-to-multipoint network*

Point-to-multipoint deployments generally involve what are known as *sectoral antennas* consisting of an arrangement of highly directional parabolic dishes distributed around a central pole. Each dish defines a sector, an area where frequencies can be reused. Generally the higher the frequency, the more sectors can be created and the higher the degree of frequency reuse. Sectors can also be formed by using what are known as *phased array antennas* where a number of omnidirectional pole antennas are grouped together and beams are formed by varying the phase relationships of their outputs. The phase relationships are electronically manipulated, and, in the case of adaptive beam forming or "smart antennas," the system will adjust beam width and direction on the fly to provide the best signal to a subscriber. Adaptive beam forming antennas, discussed at great length in Chapter 4, are likely to become the predominant technology at some point but are still quite expensive today.

Point-to-Point

Simple point-to-point links are the legacy architecture and arguably are not networks at all since they do not permit any-to-any connectivity. This is the architecture that has been most used in the upper microwave regions within the last few years, though it has a long prior history in telecommunications.

For years and even decades, microwave operators have set up isolated links for LAN extension, for connecting remote sites with the PSTN, and for cellular backhaul. Indeed it is often forgotten today that microwave point-to-point links carried a large fraction of long-distance telephone and data traffic prior to the introduction of fiber-optic networks in the 1980s.

Today point-to-point backhaul and point-to-point access for business high-rises comprise a solid but slowly declining business, one that represents a poor utilization of network resources. Simply put, a base station should serve a multitude of customers if it is to pay for itself. Where point-to-point connections are indicated by the nature of the equipment, as in the 60GHz band, then the best choice is to construct a sort of quasi-point-to-multipoint network topology by placing multiple radios at the same site and sharing one router or switch among all of them.

Point-to-Consecutive Point

Point-to-consecutive-point wireless deployments have been rather rare, but they are arguably a better way to use millimeter microwave equipment than point-to-multipoint because they generally allow the network operator to reach every potential customer in a given market. Often referred to as *wireless rings*, point-to-point networks need not form a perfect circle but can describe a zigzag pattern. As long as the signal path ultimately returns to its point of origin and forms a closed loop, the network will qualify as logical ring. One now-defunct company, Triton Networks, manufactured a complete wireless ring solution at 28GHz–31GHz and 38GHz with SONET restoration and the wireless equivalent of a SONET add-drop multiplexer at every node, but other manufacturers have not used this concept. Today the network operator wanting to adopt such an architecture with WiMAX equipment would have to jury-rig a solution using routers and would likely incur considerable expense in doing so. Optionally, that operator could use non-WiMAX point-to-point microwave equipment designed to support rings. Another option would be to design the system as large Ethernet LAN, but the problem there is ensuring fairness since in an Ethernet the user closest to the switch enjoys a great advantage in contending for bandwidth. Additionally, a considerable amount of equipment is available for a network operator wanting to construct a fiber ring, but in the wireless realm a ring is almost an aberration. The notion of a wireless ring still has merit, but today network operators wanting to implement that architecture are on their own. Figure 4-4 shows a point-to-consecutive-point deployment.

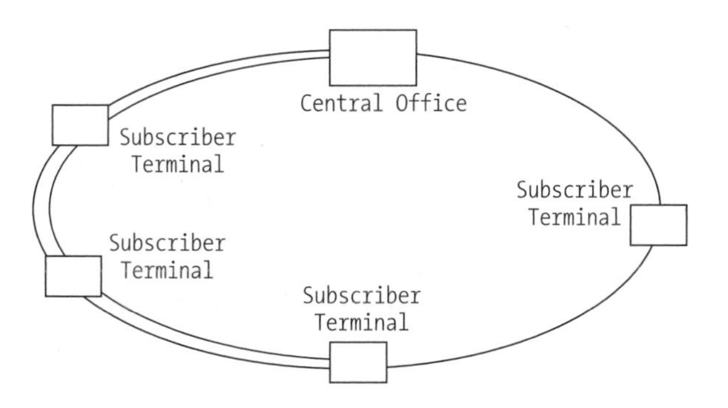

Figure 4-4. *A point-to-consecutive-point deployment*

Mesh

Finally we will consider the mesh, the least utilized network architecture to date, but by some accounts the most promising.

A mesh substandard is included in 802.16a based largely on the technology Nokia Corporation purchased from Rooftop Communications. None of the current proponents of mesh networks for broadband wireless deployments follow that substandard.

A mesh would appear to be the most expensive architecture to build—though manufacturers of mesh equipment dispute this—because each node requires a router. It also appears to be the most robust because each node has multiple pathways available to it. Since deployments have been so few, it is difficult to ascertain the real business case for mesh architectures, and it is equally difficult to determine when a mesh would be indicated over point-to-multipoint. The argument in the past in favor of the mesh was that individual base stations in a point-to-multipoint architecture could not access all potential customers because of line-of-sight limitations, but with the introduction of NLOS equipment that contention appears considerably less persuasive today. On the other hand, a mesh almost eliminates the need for backhaul, which, in many cases, is the biggest cost in setting up a wireless broadband network.

Conceptually the mesh architecture is intriguing and, if one takes the concept to its logical extreme, suggests a fundamentally different way of structuring a communications system, not just on the access level but at all levels. A pure mesh—sometimes called a *fractal mesh* because its gross structure resembles its fine structure—is essentially nonhierarchical; in other words, it forms a homogenous organization where distinctions between the LAN, the metro, and the long haul disappear as well as distinctions between subscriber terminal and network access point. Since network intelligence is distributed through the network and is self-organizing, there is no need for base stations, central offices, or large routers or switches to manage the network. Billing and provisioning still have to be centrally managed, assuming that the network is a for-profit enterprise and not a cooperative venture, but the basic hub-and-spokes or tree-and-branches structure typifying all traditional communications networks is no longer present and plays no part in the distribution of information through the network for the execution of management and system support functions. Figure 4-5 shows a mesh architecture.

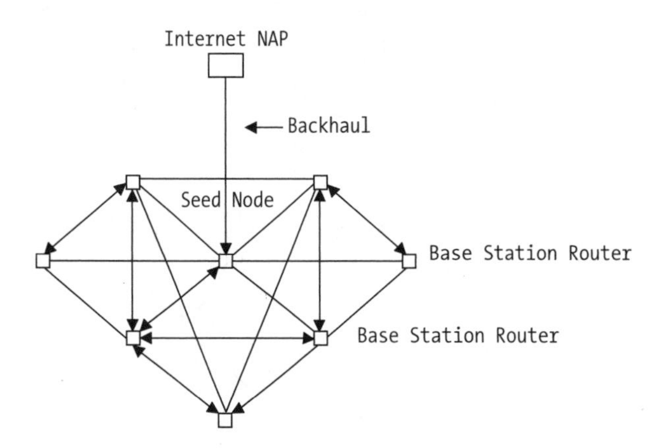

Figure 4-5. *A mesh architecture*

Types of Meshes

Mesh architectures take two basic forms: the switched mesh and the routed mesh. The second predominates today.

In a switched mesh a more or less fixed pathway through the network is established for each node, and each packet will follow the same path through the course of a transmission. In the event of a severe degradation in the link, the subscriber terminal will renegotiate routing arrangements with its fellows and establish a new pathway through the network, but this function will be more in the nature of a restoration mechanism than a means of managing traffic. This configuration is also known as an infrastructure mesh.

A routed mesh, on the other hand, will have active, intelligent nodes, which continually evaluate link conditions and establish new paths on the packet level as warranted. As with the core routers used in the public Internet, mesh nodes will weigh link performance across a number of parameters, including throughput speed, traffic density, latency, packet loss, jitter, and noise, and will make determinations based on the requirements of the individual transmission. Another term for this is the ad hoc mesh.

Routed mesh networks take a number of forms that tend to represent points along continua rather than rigid distinctions. At one extreme, the *all-knowing mesh* is where every node sees every other node and maintains complete routing tables. At the other extreme lie those meshes where each node knows only its immediate neighbors. Intermediate forms are possible as well where a single node may know a segment of the network extending beyond adjacent nodes but not the entire universe of nodes.

All-knowing meshes, by providing each node with maximum information on the condition of the network, enable each node to plot routes that best meet the requirements of the transmission. On the other hand, the memory and processing power required to store routing tables for the entire network and run global routing algorithms drives up the cost and complexity of each subscriber terminal. And because each node must poll every other node, the network overhead will be higher as well, increasingly so as the number of nodes increases.

Another pair of distinctions concerns the overall behavior of the nodes in performing route computations. Some meshes are entirely reactive, only polling their neighbors when a packet is to be sent. Others are proactive, looking at the network even when they are not transmitting data payloads. Here again the distinctions are not hard and fast, and various degrees of reactivity and proactivity are possible within the same network. Obviously, the more network diagnosis the node is doing, the less bandwidth is available for data.

The routing algorithms utilized by the surviving equipment manufacturers are all proprietary and extensively optimized for the wireless metropolitan area network (MAN) environment. They tend to diverge sharply from the routing protocols used in wireline core and edge routers. The maturity of wireless mesh routing schemes is a matter of considerable debate, and the lack of mass deployments forces the network administrator to depend on computer simulations for determining the appropriateness of any given mesh approach for a given market. Unfortunately, according to the relatively few engineers with extensive experience in designing and configuring meshes, simulations are often inadequate and misleading.

Routed mesh equipment is obviously more complicated to design and execute than is its switched mesh counterpart, but its ability to adapt to changing network conditions is greater. Today the majority of surviving mesh vendors are offering routed mesh solutions, perhaps because initial wireless mesh research was done in support of ad hoc military deployments where speed of deployment, noncritical placement, and overall network robustness were the primary considerations and where the expense of using multiple routers was not an overriding concern.

Mesh Advantages

The arguments advanced by current manufacturers in support of the mesh topology are several.

First, total cost of infrastructure is purportedly far less per a given coverage area. Mesh advocates claim that the cost premium associated with a routing capability in the subscriber terminal dwindles to insignificance with mass production and claim that the elimination of base stations with their right-of-way costs and their costly, limited production hardware, as well as their redundancy and ruggedization requirements, will save the network operator a substantial portion of his total equipment budget. Add to this the huge reduction in backhaul costs, and the supposedly less-critical siting requirements for individual nodes, and the cost argument seems quite persuasive. One vendor, California-based Tropos Networks, which does not support 802.16, quotes a $20,000 to $50,000 price per square mile for the total number of terminals necessary to provide pervasive coverage.

Second, coverage is supposed to exceed that afforded by point-to-multipoint networks, even those equipped with smart antennas. This claim appears plausible because of the routing diversity inherent in a mesh architecture, but the degree of routing diversity naturally varies according to the number and distribution of nodes in the network. A network with only a handful of nodes does not fully realize the advantages of the architecture. Metcalfe's law, which says that a network's value increases by the square of the number of nodes, is certainly true of a wireless mesh.

Third, the siting of the subscriber terminal is significantly less critical than is the case with a point-to-multipoint network because if the node cannot establish a good link with one adjacent node, it can probably do so with another. Some mesh advocates go so far as to claim that the network operator can largely dispense with site surveys and that self-installation on the part of the subscriber can become the norm. Again, I see merit to this argument, but I point out that here also the strength of the argument increases with the density of network nodes. A mesh with only a few nodes simply cannot have a lot of diverse routing paths for each of them.

Some of the more adventurous companies promoting mesh networks are New Zealand's IndraNet, Massachusetts-based Ember Corporation, and Florida-based MeshNetworks (now part of Motorola). They claim that meshes will bring about fundamentally new service models where wireless connectivity will become completely ubiquitous. The mesh itself will be used to support massively parallel grid computing where many nodes in the network will participate not only in routing decisions but in analyzing other networks such as transportation systems, the power grid, and security systems, to name just a few. Such a fully realized mesh network will also exhibit advanced storage capabilities and will serve to host a tremendous multitude of applications as well. IndraNet principals perhaps go furthest in touting the transformational capabilities of wireless mesh networks, suggesting that they will bring cheap pervasive voice, data, and video connectivity to every corner of the world and will exert revolutionary effects on everyday life almost everywhere.

It is difficult not to react sympathetically to the idealism animating such speculations (because who would not want the benefits of advanced communications made available to all people?), but network operators striving simply to establish a profitable business in a given market are still forced to ask how well a mesh approach will serve the present needs of establishing an initial presence in a market and ultimately reaching the key customers necessary to sustain the business.

Absent any significant number of successes in actual deployments, the answer to that question is elusive, but I offer the following observations.

Distributed intelligence and parallel computing are no longer radical ideas. Major information technology companies, including Sun Microsystems and IBM, are firmly behind the notion of grid computing, and indeed the fastest computers made today are no longer expensive, one-off supercomputers but instead vast assemblages of conventional CPUs linked by software. That a computing grid could manage a metropolitan network is by no means inconceivable, and that it could manage it more cheaply and effectively than a central office rack of big iron switches and servers is possible though not proven.

Mesh deployments are said to have been used successfully by the U.S. armed forces in the field, but the federal government has divulged little information on the scope of such deployments and the traffic densities occurring within them. Nokia sold its Rooftop Communications infrastructure equipment to a handful of service providers, providing some commercial field experience, but the Finnish company has since ceased production of the radios. Finally, several service providers have set up wholly proprietary mesh architectures of which the now-defunct Metricom Ricochet network with more than 30,000 subscribers was by far the largest. Whether these constitute real proof of concept is debatable because none of the service providers has survived.

Mesh Networks: The Counterarguments

While freely acknowledging the revolutionary potential of the mesh, you will see the following challenges facing those who utilize mesh architectures in full-scale metro deployments:

Developing routing algorithms that maximize network efficiency and minimize congestion, unacceptable latencies, and "router flaps" where highly inefficient routing paths are inadvertently selected: Packet routers function imperfectly in public networks today and do not provide the same kind of wholly predictable network performance as do traditional circuit switches. Indeed, popular protocols for improving router predictability, such as Multiprotocol Label Switching (MPLS), DiffServ, and Reservation Protocol (RSVP), make the operation of the router notably more switchlike and set up fixed paths through the Internet that essentially ignore the basic mesh structure formed by the cross-connections among the major peering points.

Securing the network against hacks and malicious code: If the basic neurons of the network are exposed, which they certainly are in any peer-to-peer arrangement, then it is difficult to protect the core management software. Advocates of grid computing have been preoccupied with network security, and rightly so, but really satisfactory solutions have not been developed to date. Essentially closed systems such as class 5 circuit telephone switches, ATM switches, and optical switches have generally been highly resistant to intrusion and compromise whereas IP routers are almost routinely subject to successful attacks. There are those who argue that nonhierarchical systems will be more resistant to attacks, and the effects of such attacks will be localized rather than catastrophic, but an absolutely convincing case has yet to be made for this position.

Minimizing hops: The longer the chain of wireless routers going back to the Internet access point, the less bandwidth available to any given node for transmitting its own data, and this constraint ultimately limits the size of a mesh network. If a mesh network grows sufficiently large, nodes on the periphery will have to go through multiple intervening nodes, which severely limits throughput for any one node. This deficiency is exacerbated by the inability of most radios to transmit while they are receiving.

Developing standards: Single-vendor solutions are the norm in mesh networking today, a state of affairs that, ironically, militates against the kind of pervasive deployments mesh advocates envision. Currently little comprehensive standards work is under way in the area of wireless meshes, and no major telecommunications infrastructure manufacturer is directly involved in mesh projects.

Providing a strong first-mover advantage and a real ownership of the network for the network operator initiating the mesh: If the intelligence is in the subscriber terminal, there is really nothing to stop subscribers from forming loose cooperatives and dispensing with the middle man, so to speak, or else reducing the middle man to a role of simply negotiating and maintaining a link with an Internet Point of Presence. Mesh equipment, inasmuch as it is nonhierarchical and self-organizing, inherently favors ad hoc, cooperative associations of nodes, and some of the vendors have acknowledged as much. It is rather difficult to ascertain what the precise role of the network operator would be in a truly self-organizing mesh other than serving as a sole conduit for a monopoly infrastructure product, and I cannot see anyone fulfilling such a role indefinitely. It well may be the case that a mesh of meshes on a national scale is infeasible and that local meshes simply cannot coalesce into an all-encompassing canopy, but if they can and do, the function of the service provider must change radically. Technology pundit George Gilder speculated on the properties that a nationwide fiber-based mesh network might exhibit in his book *Life After Television* (W. W. Norton & Company, 1994), and he assumed that the service provider's role would consist merely of owning and maintaining the actual fiber plant, but in a wireless mesh there are no actual links to own. The service provider's place in such a scheme is obviously at issue.

A final challenge is the attitude of entrenched wireline incumbents in the event that wireless meshes begin to win acceptance. One could foresee such incumbents seeking some sort of legislative or regulatory relief to preserve the de facto monopoly they enjoy today. And in the United States one could easily see them prevailing against the interests of the public.

All of this does not preclude a network operator soliciting a bid or a proposal from a wireless mesh manufacturer. All of the claims made by such manufacturers may yet be realized, and meshes may eventually become the norm rather than the exception. I think, however, that meshes, because they are relatively untried, represent a higher risk option than more traditional architectures.

Performing Site Surveys and Determining Link Budgets

A *site survey* is simply a determination of the extent of the interference levels in the frequencies of interest and of the severity of obstructions and multipath at the precise locations in which one wants to install terminals. Such a survey enables the network operator to determine if unacceptable amounts of interference are present at various locales such that adequate signal quality cannot be obtained, as well as the extent to which blockages restrict blanket coverage in a given area. Ideally, the site survey should be an ongoing process and should encompass every link in the network. In practice, something considerably less extensive will be performed because of the prohibitive cost of deploying technicians at every location.

Site surveys are performed manually and painstakingly by taking innumerable measurements with a radio spectrum analyzer and eventually with an actual radio transceiver. It is grunt work, but a certain minimum surveying is required for informed network planning. It is also one of the first steps a network operator should take, and if unlicensed spectrum is to be utilized, it should be perhaps the second step after a determination has been made as to what spectrum is available.

In theory, a site survey should encompass every location where either an access point or a subscriber terminal may be situated, but of course the time and expense required to achieve that theoretical ideal are enormous in the case of a large metropolitan network. Therefore, almost inevitably network operators will have to satisfy themselves with a less-than-thorough accounting of the wireless environment in which they will be operating. At the least, site surveys should be done for all base stations and for all subscriber terminals serving multiple customers such as terminals at business parks and high-rises, and multi-tenant residential buildings.

In the case of NLOS radios and mesh architectures, the importance of the site survey may be lessened. After all, the intent of both approaches is to foster self-installation, and that virtually precludes a site survey except perhaps through some simplified software utility included with the subscriber terminal. I have already touched upon meshes and have indicated the many uncertainties associated with their deployment and the plethora of paths over which an individual transceiver may transmit. Obviously a comprehensive site survey within a mesh network would be nearly overwhelming. Chapter 5 covers NLOS approaches extensively, including issues of terminal siting, but here you should observe that NLOS self-installed equipment is generally placed indoors, which greatly reduces the maximum permissible distance from the base station and, if generally used throughout the network, may require an increase in the number of base stations over that of a deployment characterized by professionally installed outdoor antennas. Nevertheless, some will argue that line of sight always entails more base stations in the end in order to get around blockage.

If an extensive number of indoor installations is under consideration, the network operator may be forced to weigh the expense of more base stations relative to more truck rolls. I have already mentioned the cost of truck rolls. A truck roll involving not only installation work but a complete site survey may take the better part of a day and may cost the network operator hundreds of dollars. If such truck rolls are required for most customers within a large metropolitan network, and the customer population eventually grows into the thousands, they will represent a huge operational expense, one made worse if the network experiences a lot of churn. On the other hand, base stations can entail considerable up-front capital expenditures and significant recurrent costs. Within the lower microwave regions a base station may carry an up-front cost as high as $10,000, a backhaul cost of $1,000 per month, and a site lease cost of another $1,000 per month (these are typical figures, but by no means benchmarks) for a total of $34,000 during the first year of operation, not counting utilities and allocation of staff time. On the other hand, 100 truck rolls performed within a single cell could exceed the total expenditure on one base station.

I prefer NLOS solutions because they enable the network operator to reach customers who would otherwise be inaccessible as well as reduce the number of site surveys that must be completed. One should understand that NLOS is not a panacea, however.

To return to the discussion of actual site survey procedures, note that the person conducting the site survey is seeking three kinds of information. First, he is looking at the amount of interference in a given band and adjacent bands either from existing users or from accidental interferers. Second, he is attempting to determine what the signal strength will be at the antenna at the site in question for signals emanating from various other points in the network. Finally, he is looking at the integrity of the data stream and such factors as bit error rate, jitter, latency, and throughput rate.

Although currently available instrumentation is precise, there is still apt to be a certain amount of guesswork in doing the survey because one can almost never determine the concentration of customers in given areas covered by a base station, and course interference levels can change over time, sometimes markedly. The initial site survey does provide the network operator with baseline information, however, and as such it is an indispensable part of the planning process.

A network operator can elect to use internal staff to perform the site survey or use outside consultants. In either case prior experience is highly desirable. At a minimum the individuals performing the survey must be familiar with the basic concepts discussed in the next section.

While the survey is under way, all observations and readings should be carefully and completely recorded, and the final report should be kept in a secure place where it can be consulted by responsible individuals in the organization as the need arises. The report should indicate the exact locations of all network nodes, including elevations. Using the Global Positioning System (GPS) coordinates to record positions is an excellent idea.

Determining Line of Sight and Computing Fresnel Zones

The site survey begins with visual observations, and the first task is to attempt to establish clear line of sight from the subscriber location to the base station; the assumption here is that right of way has already been granted for antenna placement in both locations. The process is quite straightforward and involves nothing more than standing at the level of the transceiver antenna in one of the locations and training a pair of binoculars on the other antenna site. If a clear path lies between the two points with several yards separating the nearest obstruction from the imaginary line connecting the two points, then one can assume that line-of-sight preconditions have been met.

Another method available in the United States, though not always elsewhere, is to purchase three-dimensional aerial maps of the area one wants to serve. These provide elevations of all buildings and natural obstructions such as hills and bluffs. Arizona-based AirPhotoUSA is a vendor that has served many wireless operators.

If obstructions directly cross the imaginary line, then one obviously lacks clear line of sight. If obstructions nearly cross that line without quite impinging upon it, then one must proceed to the next phase, the calculation of what is known as the *Fresnel zone.*

By strict definition, Fresnel zones consist of an infinite series of concentric rings surrounding the nodal point of transmission, with each ring defined by the phase relationship between the main beam of the transmitter and the two dominant side lobes (these terms are explained shortly). The phase fluctuates from one zone to the next so that side lobe reflections are sometimes in phase with the main beam and sometimes antiphase to it.

So how do such side lobes arise? Directional antennas of the sort generally employed in broadband wireless networks tend to focus energy into narrow beams, but they never produce just one beam. Rather, they produce a dominant central beam and two or more side lobes that contain less energy than the main beam. They also produce low-energy rear beams. Depending on the polarization of the transmitter, the so-called side lobes may be above or below the main beam or on either side of it.

Now, even when the direct signal has a clear path to the receiver antenna, these side lobes may encounter obstructions since they are offset from the main beam and describe different propagation paths. When they do strike obstructions, they will be at least partially reflected, and the reflections may impinge upon the main lobe, either reinforcing or canceling the signal depending on the phase relationship of the two.

Here an element of confusion enters the discussion because field technicians tend to employ a different definition of Fresnel zones. By this second definition, a Fresnel zone refers not to a concentric ring but to an elliptical volume surrounding the imaginary line of sight and representing a region where the presence of physical obstructions will set up strong side lobe reflections that may disturb the main beam.

The reflected signal may, as you have seen, be in phase or antiphase to the main beam, and thus it may reinforce it or partially cancel it. One may think that reinforcements would be preferable to cancellations, but in fact both are undesirable, especially so within the first Fresnel zone (following the earlier definition) where strong early reflections of the side lobes occur. The network operator simply has to plot a path that is free from major obstructions.

The distances involved in Fresnel zones are frequency dependent and are also a function of the radiation patterns of the antenna used. It is best to consult with the manufacturer of the subscriber terminals and base station equipment to determine the extent of the area above and below line of sight that must be kept clear of obstructions. Most commercial programs for cell site location include subprograms for Fresnel zone calculation, and Proxim, a leading manufacturer of broadband wireless equipment, makes a well-regarded stand-alone program, so one need not despair if one's math skills are wanting.

If obstructions do stand between the base station and a valuable subscriber site such as an office building or residence with multiple subscribers, one remedy may be simply to raise the base station antenna on a mast at an elevation where it is well above any obstructions. Such a tactic must not be regarded as a perfect solution, however, because an antenna that is too high will not be able to reach subscribers in the immediate area of the antenna.

It should be further noted that not all obstructions are equivalent and that considerable differences may exist among obstructions of the same general type.

A single tree for instance may impose around 15dB to 20dB of signal loss depending on type and size. A grove of trees may up that figure to 30dB. A building may represent a total loss of 30dB while a low hill could exceed 40dB. A truly interesting situation occurs when trees sway in the wind. Momentary variations in loss may exceed 10dB, and generally one must design around the worst case.

Bear in mind that the presence of obstructions within the first Fresnel zone does not preclude the establishment of an airlink. It just means that more transmit power will be needed to achieve the same signal integrity as in an unobstructed path.

RF Analysis

Once the site surveyor has been able to establish whether an airlink is obstructed, and how severely, if obstructions exist, the time has come to evaluate the RF environment for that airlink and determine whether it is ultimately useable. This happens with either a dedicated spectrum analyzer or a specialized software program run over the actual data radio equipment.

All the major instrumentation companies—including HP/Agilent, Anritsu, Wiltek, and Rohde & Schwartz—make dedicated spectrum analyzers, and most make handheld models that can be managed by an individual obliged to mount a radio tower. These are precision instruments capable of wideband tuning and of detecting activities in nearby bands that may impact transmissions in the bands selected by the network operator. While not inexpensive, they are designed for outdoor use in adverse weather conditions and will serve the operator well over years of installation work. And if operators are unwilling to purchase such a device, they can rent them, though I strongly recommend purchasing one since in a growing network it will be in more or less continuous use. Also, if the adaptive array antennas are used, a special type of spectrum analyzer known as a *deep memory waveform generator* will be required. Agilent is the principal manufacturer of such devices.

Increasingly, broadband radios include built-in spectrum analyzers, reducing the need for specialized instruments. In all cases the radio itself will be far less portable, however.

Unlike the determination of line of sight, the RF examination should be done at both ends of the link, since there is no way of determining the presence of interference away from the point where it is experienced. Make sure that readings are taken at the precise location where the antenna is mounted. It does little good to make measurements from street level if the antenna is to reside on the roof of a six-story building. It is also advisable to take a number of measurements over the course of several days since interference is often intermittent. As a network operator, you will be obliged to offer high performance over your network at all times, so you must be aware of any potential problem.

In effect, each individual link requires its own site survey, though network operators striving to launch a network will most likely confine themselves to a round of surveys encompassing the most promising customer sites. It is also a good idea to get a general reading of the entire metropolitan area where one has chosen to situate one's network through random spot surveys. If, for instance, one finds that the city center is already densely occupied with private WLANs utilizing 2.4GHz, then one may want to consider another unlicensed band. In performing such a general reading, it is a good idea to rent a crane or a vehicle equipped with a hydraulic "cherry picker" so that readings can be taken at heights where an antenna may ordinarily be installed.

The purpose of the spectral analysis phase of the survey is simply to measure the level of interference at the antenna site and nothing else. The spectral analysis does not and cannot tell one what the signal strength at the antenna will be.

At the point when network operators have completed a spectral analysis of the two ends of the airlink in question, they must then come to some conclusions about the integrity of that airlink and its suitability for providing networking services. Such a determination involves some rather complex technical issues.

The mere presence of interference does not rule out using the link for data communications. All commercial radios tolerate some degree of interference, and, in the case of radios operating in unlicensed band, the presence of significant background interference is assumed. Interference becomes critical when the level of interference approaches that of the signal.

Individual radio designs differ in their ability to reject interference within the same channel and adjacent channels as do different modulation techniques in general. For instance, frequency hopping confers an especially high immunity to interference because the signal becomes in effect a moving target as it hops from one channel to another in a synchronized pattern. Coarse modulation techniques such as 16 quadrature amplitude modulation (QUAM) also provide superior rejection of noise and interference. Unfortunately, however, those techniques that render the airlink more robust in the presence of interference may also limit the speed of the connection or reduce the electrical efficiency of the radio.

The issue is further complicated by the fact that the TCP transport protocol, mandated in the case of 802.16, has provisions for resending packets in the event of interference-induced losses. So long as the interference has some degree of intermittency—that is, it falls below the level of the signal sufficiently often to permit some packets to go through—a transmission can be maintained even in the face of heavy interference, though at considerable cost of speed and reduction of service quality. The question then becomes, what is acceptable to the customer?

One should also understand that the implications of network nodes operating in a highly interference prone environment go beyond the individual customer. To the extent that an individual radio is doing a lot of retransmitting, that radio itself is elevating the general noise floor in the area and causing other radios to retransmit more frequently, which in turn further degrades the performance of the first radio in a vicious circle. In a sense, radio interference is a bit like acoustic feedback in a public address system. One can get only so much level out of the system before escalating feedback buries the signal.

The equipment manufacturer that one has chosen to work with should be able to provide information about levels of the interference in terms of decibels below the reference level of the signal that the radio can tolerate while maintaining a good data link with few resends. If interference regularly exceeds the recommended level, then one must consider abandoning that link.

The temporary installation of radios at both ends of the link will provide the final determination as to whether the link is useable. The radio's own diagnostic software should permit the surveyor to measure signal strength and to evaluate the availability of the link. At this point in the process experimentation with the placement of the antenna is entirely in order, and the operator may find that adjustment on the scale of centimeters can make the difference between a good airlink and its opposite.

Analyzing the Data Stream

Once the RF analysis has been completed, the network operator will want to proceed with a temporary installation of terminals at either end of the link and proceed with an evaluation of the data link. Here one should measure the data throughput, latency, packet loss, and number of retransmissions. Most manufacturers of wireless broadband equipment include software utilities that support such diagnostics. Since the radio terminals are already in place, and there is nothing to be moved physically, the data link evaluation is a much more straightforward process.

Establishing the Link Budget

The final step in the whole process is establishing what are known as *link budgets* for the individual airlinks. This term describes a series of calculations for determining a reliable connection between two points, and it is just that—a budget encompassing the signal gain provided by the antenna and the external RF amplifier, if one is used; the losses imposed by the normal attenuation of the signal over distance; the losses from varying atmospheric effects; and the losses from various equipment-related issues at the receiving end such as ohmic losses in the connector cables. Also figuring in the calculations are the gain of the receiver's antenna and the sensitivity of the receiver, that is, the threshold signal level required for the receiver to achieve a specified output level at a specified signal-to-noise ratio. Normally the network operator will want to establish a margin of at least 40dB above the sensitivity threshold of the receiver; this is termed *fade margin.* This margin ensures that the transmission will not be interrupted by temporary increases in interference or by unusually adverse atmospheric conditions.

Note that fade margin is directly related to availability (the percentage of time the network is down). Conventionally, each 10dB of margin translates into an order of magnitude increase in availability. Thus, a radio link with a fade margin of 50dB will have five nines (99.999 percent) availability, which is the standard metric for wireline systems.

Link budget computations differ considerably depending on the band in which the network is operating. Atmospheric attenuation, for instance, is relatively unimportant in the lower microwave region but hugely important in the millimeter microwave region. In the first instance it can generally be left out of the calculations, and in the second it is ignored at the operator's peril.

In calculating the link, budget network operators have one main variable with which to work—namely, the transmit power of the nodes. By increasing power they also increase the fade margin and rise above the background interference. But they also increase the interference seen in other links, and they soon face legal limits as well. Broadband access radios are generally limited to a few watts maximum just about everywhere, and for the unlicensed bands the figure commonly is 1 watt.

Given the diversity of terminal designs in the marketplace and the changing parameters over frequency, a complete listing of all relevant calculations is impractical. As in many other areas of network design, manufacturer support is essential here.

Again, formal link budgets are generally established only for the most critical airlinks, those serving multiple high-value customers or providing base station backhaul. The cost in technician time of optimizing the link for each subscriber to basic access services is just too great. The following is an example of a link budget table, courtesy of Alvarion:

$$S_P = P_T+G_B+G_T-P_L-R_A-L_S-FM-Sensitivity$$

where the following is true:

S_P: Spare (available margin) over sensitivity

P_T: Transmitted power

G_B: Base station antenna gain

G_T: Subscriber unit antenna gain

P_L: Path loss (using a chose propagation model)

FM: Fade margin (as required for quality of service)

R_A: Rain attenuation

L_S: System losses (cables attenuation, implementation losses, antenna gain reduction)

Examining the Equipment Selection Process

Traditional telephone carriers go through a laborious formal process of qualifying and selecting equipment. First a request for proposals will be issued to vendors deemed suitable, and then the equipment submitted will be subject to an exhaustive testing process. Finally, a bidding process will be initiated, and the carrier will make its selection.

A wireless broadband service provider, particularly a startup, may lack the lab facilities and staffing to do anything nearly as thoroughgoing, but a careful systematic process is still indicated and indeed is essential.

Generalities

The process of equipment selection is as much about assessing the manufacturer as the actual equipment. Wireless installations are seldom "slam dunks," and manufacturer assistance and support are invaluable. One wants to choose a manufacturer who will take an active part in at least the early stages of the installation and that can provide support if difficulties arise later. Obviously, to fill such a role, the manufacturer must avoid bankruptcy, and the large number of business failures among broadband radio manufacturers should incline network operators to scrutinize the stability of the vendor closely. Unfortunately, the solidity of the manufacturer does not necessarily correlate with the quality or performance of the equipment, so one must look at the vendor's business practices as much as at its products.

All RF equipment should be thoroughly measured as to compliance with performance specifications and then used in actual links to determine performance in actual field conditions. During such evaluation, pilot links should not be operated as commercial services. It is entirely illegitimate to do product testing at the customer's expense, and the ill will that can result from such a practice can subsequently jeopardize the prospects of the wireless service.

Field testing of equipment in semipermanent links should be designed to expose the equipment to as variable an environment as possible in order to assess the effects of high levels of interference, severe multipath, network congestion, large obstructions, and so on. The key attribute of carrier-grade equipment is consistency of performance.

Be wary of claims of such capabilities being made available in a later product iteration, especially if the development of that capability is likely to challenge the resources of the manufacturer.

Look to easy upgrade paths when considering equipment. Will a particular network element be able to grow with the network, and, if so, what is the cost of expanding port counts or adding new types of services?

Never assume full interoperability based on a company's professed adherence to a standard, even with certification. Always determine interoperability based on actual field trials, and never purchase equipment that is to interoperate with existing infrastructure absent such a determination.

If equipment is to reside outdoors, make certain it is rugged and complies with industry standards for weather resistance. In the past, many manufacturers have paid insufficient attention to the design of the case, and the equipment has underperformed as a result.

Finally, contact every relevant manufacturer, at least at the inception of the network, and invite competitive bids. Few wireless broadband operators have sufficient sales that they can afford not to be accommodating in respect to price. Wireless broadband today is a buyer's market.

Service-Related Specifics

The correct equipment choice for network operators is equipment that best supports their service offerings. Chapter 6 explains in detail how equipment supports services and indicates the equipment complement required for various offerings. The 802.16 standard provides a full feature set that will enable a considerable range of service offerings, but for services such as streaming multimedia, conferencing, and voice telephony, special provisions not present in every system will be necessary. Do not assume that any given piece of equipment can be jury-rigged to provide every conceivable service.

One should also make certain that the radios are provided with all necessary interfaces for other types of networks with which the broadband wireless network might communicate. Many, though not all, radios have fiber interfaces, for instance, and if fiber is to be used for backhaul, better to have interfaces in place rather than having to set up outboard optoelectronic converters.

Another technical consideration is the number of RF ports on the base station controller. If the operator wants to operate on multiple bands, that many RF ports will be required.

Routing and switching requirements should also be determined before any equipment contracts are signed. It is certainly possible to use a base station controller with an outboard router or switch, but it is generally a lot cheaper to buy an integrated network element that does it all.

Lastly, investigate how the component's element management system will fit in with the overall operations support systems (OSS) software suite utilized in the network. Few startups or competitive service providers can command deep expertise in network management software, and if off-the-shelf programs can be bolted together and made to work reasonably well, that is desired. What one wants to avoid is having to hire consultants and trying to harmonize the software for a single element with an overall system with which it is fundamentally incompatible.

Integrating Wireless with Wireline Infrastructure

Wireless infrastructure should always be regarded as an option, not as a cause to pursue. Sometimes it is a good option, sometimes it is not, and sometimes the best choice that a network operator can make is to put wireless links in some parts of the network but not in others. But be aware that when some portion of the network is wireless and some part wireline, integration problems may occur.

One fact cannot be emphasized enough in this regard, namely, that all the commonly used networking protocols, including SONET/SDH, Ethernet, IP, MPLS, RPR, frame relay, and ATM, were devised for use over wireline networks—networks where high availability was a given. In fact, most of the more recent protocols were explicitly designed for use over optical fiber where packet loss may be presumed to be negligible. None of these protocols work well over wireless airlinks in native form, and wireless standards such as 802.11 and 802.16 all incorporate invisible layers of middleware that adapt the wireline protocol to the exigencies of the RF environment.

When a network operator attempts to use a standards-based airlink to carry a protocol for which no appropriate middleware is present, performance is apt to be degraded, or the link may even fail altogether. An example would be attempting to carry SONET or ATM traffic over an 802.11 link. Both standards happen to lack mechanisms for retransmission, and both depend upon precise clocks to coordinate data flows. Furthermore, ATM has procedures for negotiating sessions that depend on the maintenance of precise timing and are intolerant of interruptions in the signal. Obviously, these protocols will not function correctly over an intermittent airlink absent appropriate middleware, and in the case of SONET, the restoration mechanism may even be invoked.

Further problems may be anticipated when a multiservice switching platform sits at the core of the network in question and that platform has proprietary enhancements placed on top of the familiar transport protocols. One simply has no way of knowing how the middleware in the 802.16 airlink will cope with those ancillary protocols, but because they have been designed around a fiber connection, the likelihood is that problems will arise.

The 802.16 standard generally supports IP, Ethernet, and ATM. No support is specified for frame relay, MPLS, RSVP, DiffServ, resilient packet ring (RPR), DOCSIS, MPEG video, or any IP voice standard. Some manufacturers may support some of these protocols by means of proprietary middleware hooks, but one can never assume that.

If one owns or contemplates owning networking equipment supporting protocols other than Ethernet and IP, one should determine in advance that the wireless broadband equipment to be added to the network is compatible. Improvised solutions may be possible through the agency of a middleware developer, but one should not count on that.

Assembling the Management and Technical Teams

In planning a wireless broadband network, one must not neglect the human dimension because staffing decisions will be vital to the success of the network.

If, as is often the case in wireless startups, initial funding is limited, the temptation on the part of the entrepreneur founder is to assume the role of chief cook and bottle washer and attempt to do everything. Indeed we have seen countless instances of entrepreneurs with no prior experience in either wireless data or network management attempting to perform all the business functions involved in running a network as well as taking on the duties of chief technical officer. Sadly, this kind of overreaching almost always results in the operation's failure.

Adequate staffing is just as important as installing the right equipment. Network operation is a hands-on business, presenting constant challenges even to experienced management and technical teams. If capable individuals are not in place to perform all the vital functions involved in setting up and running a network, the operation is almost certain to founder.

In some cases, individuals with expertise in more than one function and in the early stages of network growth such multitasking may be acceptable. But as the network grows, so must staff if network performance is to meet the expectations of the subscribers.

Rather than describing specific positions, I will simply enumerate the major responsibilities in launching and running the network.

First, overall network planning is essential. This function concerns itself not only with the airlinks themselves but with the core switching and routing components required to handle network traffic, as well as the physical and software platforms required to support specific service offerings. Such planning involves basic questions about how the network will be architected to serve the needs of various groups of subscribers and how the initial deployment will support expansion.

In established carriers, whole teams of network engineers engage in such planning exercises under the leadership of network executives, but a wireless independent will rarely be in a position to hire a planning team. More than likely a single individual will have chief responsibility for developing the network architecture.

This may not be altogether a bad thing. Traditional telcos have separated those with expertise in routers and data switches from those who oversee traditional voice switches, reflecting a dichotomy that is becoming increasingly meaningless today. The people planning a modern pure packet network need to be generalists, so specific experience in one area within a large incumbent may not be the best preparation for a broadband wireless network planner.

Second, the organization needs to have on staff someone with extensive hands-on technical experience in implementing and maintaining wireless data networks. Relying on wireless consultants is simply inadequate, and hiring someone who has run large wireline data networks but never dealt with RF airlinks will not suffice either. Unfortunately, highly qualified individuals in this field are not numerous, and their scarcity is apt to be reflected in their salary demands. The individual in question need not be a degreed engineer, and in fact most inexperienced electronic engineers know next to nothing about radios, but the candidate must be thoroughly familiar with issues such as link budgets, multipath distortion, Fresnel zones, and the whole range of technical details having to do with system setup and maintenance.

Some qualified individuals may have experience in the broadcast industry, some in mobile communications and some in military radio. Whatever their prior experience, radio experts must be highly technical, have a strong theoretical background in RF engineering, and have extensive operational experience. In addition, this individual must have an in-depth understanding of wireline data networks. Since more than likely this same individual will be supervising field installation, something more than a pure design engineer will be required.

Because recruiting customers will be vitally important as the network is being launched, a seasoned customer relations manager, preferably one with specific telecommunications industry experience, will also be necessary.

A sales force will definitely be required, a particularly important contingent in respect to business customers. In time, a marketing director will also probably be required to formulate longer-term marketing and positioning strategies for the organization.

Honest and competent financial officers are absolutely essential to the operation of a startup because operating and build-out expenses have a way of quickly getting out of control in these settings. Such individuals should report directly to a board of directors.

Initial professional staffing may be lean—one hardly requires layers of management when the network itself may only have a few dozen customers—but it should be a bit ahead of the momentary requirements of the organization. A broadband access service provider has to grow, and staff must to be in place to manage the growth.

Starting a competitive access provider, wireless or otherwise, is not a nine-to-five job and depends heavily for its success on the staff's enthusiasm and dedication. Compensation policies and managerial practices should reflect that everyone is in it together and everyone is vitally necessary for the enterprise's success. An executive or senior manager accustomed to operating within an entrenched bureaucracy is not well adapted to succeeding in such an environment, and I have seen plenty of "bell heads" with much relevant experience but inappropriate attitudes stumble badly within startup service providers. Ultimately, flexibility and resourcefulness are the most important attributes in a wireless start up team member.

Estimating Operating and Capital Budgets

Many excellent books on budgeting for business enterprises have been written, so I will not attempt any extensive description of the process here. Instead I will indicate the major items that should appear in the capital and operating budgets of the wireless broadband operator.

The capital budget will include the major infrastructure equipment that will be necessary to launch the network and will be amortized over time. Often, perhaps most often, the actual radios will not constitute the biggest capital expenditures.

If the network operator elects to use licensed spectrum in part or entirely, then licensing fees may well constitute the biggest capital expenditure, and these will vary enormously according to the spectrum sought and the circumstances under which it is purchased. Participating in government auctions for first rights to newly opened spectrum can be ruinously expensive, and obtaining spectrum as a distressed asset from the unfortunate participants in prior auctions can be a steal. In general, licensed spectrum for use in public broadband networks is not too generally available, but one never knows what will turn up in a given market. It always pays to check federal records to find who has what spectrum in the locale one has chosen to operate and then to explore the availability of such spectrum.

If the network operator is attempting to function as an ISP as well as an access provider, the network operator may have to invest in large routers, black-box encryption devices for creating virtual private networks (VPNs), server farms for Web hosting and Web site mirroring, an IP voice softswitch, multimedia servers for distributing specialized content and service offerings, and various suites of element management software for running the individual devices, plus the additional servers and workstations to host such software. In addition to element management software, the network operator will need provisioning software, customer relations database software, network-attached storage, billing software, and perhaps a global OSS system to tie it all together. And because everything has to be super reliable and redundant, at least some of the operational facilities will have to be mirrored, and backup power must be available that is sufficient to run the facilities, not for hours but for days. All of this together is apt to run well into the six figures. Table 4-1 lists typical capital expenditures of wireless broadband.

Table 4-1. *Major Capital Expenditure Components, Courtesy of Alvarion*

Type of CAPEX Investment	Existing Operator: Investment Required?	Greenfield Network: Investment Required?
CPE investments	Yes	Yes
Installation and commissioning CPE	Yes	Yes
Radio/network planning	Yes (partial)	Yes
Base station investment	Yes	Yes
Roof/tower site acquisition cost	Yes (partial)	Yes
IP switching/routing network	Partial	Yes
Backhaul	Yes (Partial)	Yes
Network management and NOC	Partial	Yes
Customer care and billing	Partial	Yes

Today much networking equipment is priced on a per-port basis so the network operator can add facilities incrementally and cost effectively, but the capital expenses can still be considerable, particularly if one is attempting to emulate or surpass incumbent wireline operators. And that's not even counting purchases such as office furniture and equipment and business condominiums.

Many of the companies attempting to establish wireless broadband networks in the past focused on providing basic access to underserved rural markets and thereby avoided really heavy capital expenditures, but that strategy may be inadequate for the future. Increasingly, all successful networks will offer converged services, and those operators who will not or cannot do so are going to become irrelevant.

Obviously, the network operator will have to secure backhaul, which will either entail building one's own links, wireless or otherwise, or entail leasing capacity. Wireless links will be minimally several thousand dollars apiece apart from the cost of securing roof rights, and millimeter microwave links may run into the several tens of thousands of dollars to purchase. Free-space optics are apt to run somewhere in between.

Operating expenses will include staff payrolls, normal business expenses such as utilities, office leases, insurance, auditing, and costs more directly associated with the core mission of the business such as maintaining equipment, advertising service offerings, maintaining a fleet of vehicles for field operations, and so on. If the network is maintaining a large data center, electrical requirements may be considerable because data and telecommunications equipment is extremely power hungry. Table 4-2 lists typical operating expenditures for a broadband wireless network.

Table 4-2. *Operating Expenditures, Courtesy of Alvarion*

Type of OPEX	Existing Operator: Investment Required?	Greenfield Network: Investment Required?
Roof/tower lease	Yes (partial)	Yes
Base station and O&M	Yes (partial)	Yes
CPE O&M	Yes	Yes
Network O&M	Yes (partial)	Yes
NOC O&M	Partial	Yes
Leased line rental	Yes (partial)	Yes
Office expenses	Partial	Yes
Advertising/subscriber acquisition	Partial	Yes
Facilities	Partial	Yes

A million U.S. dollars is not an excessive sum of money just to get started in a single municipality, and funding an ambitious build in a second- or third-tier city requires a war chest of many millions. One may hope to bootstrap a wireless broadband business with the revenues from one's customers, but it is difficult to cite many examples where this has been successfully accomplished.

Countless underfunded wireless ISPs (WISPs) have come and gone in the United States over the course of the last five years, probably on the order of several thousand altogether. Taken in aggregate they demonstrate powerfully that one cannot launch a public network simply by taking out a second mortgage on one's house and expect to succeed. Instead one must secure serious investment and patient investors who do not expect sudden salvation from a timely initial public stock offering.

I cannot outline in detail here procedures for courting such investors. I will say that network operators proposing to use unlicensed spectrum generally have a far harder time attracting investment capital than those holding licenses because unlicensed spectrum itself is not valued and is not an asset that can be disposed of if the network fails. The best course to follow in securing capital is to devise a highly detailed business plan that goes into specific service offerings and the necessary expenditures to support them and that indicates convincingly how they will be delivered profitably over the wireless infrastructure the network operator has selected. Avoid speculative leaps about "paradigm shifts" in subscriber practices or unfounded projections concerning demand for bandwidth, and try to base as much of the plan as possible on demonstrable fact. The plan must be flexible simply because high-speed access is such a volatile business, but at the same time it must embody clear objectives, not just in terms of achieving revenue goals but in building a business that will attract and retain desirable customers.

Examining Return on Investment for Wireless Broadband Networks

Calculating return on investment (ROI) for capital improvements within an established company can be difficult because of uncertainties in attributing changes in revenue or reductions in expenses to any single change in the organization. With a wireless startup the matter is otherwise because one is simply balancing the total expenditures required to launch and operate the network against total revenues derived from it.

This being the case, calculating ROI should be fairly straightforward, but it rarely is with wireless broadband service providers—or with any other competitive carrier for that matter. One cannot know for a certainty what the take rate for the new service will be, one cannot know what churn will be, and one cannot know what installation costs will be, though one can make some fairly informed estimates there.

Consistently, in the past, competitive service providers have grossly underestimated the cost of doing business and have relied on the capital markets to sustain them while they attempted to build out their networks. Such a business strategy is entirely untenable today and should not even be contemplated. Instead of aiming to build a pervasive footprint in one's designated markets immediately and then sorting out the details of actually running the network, one should ensure that each link functions perfectly in the beginning, that customers' needs are served in a timely manner, and that cost-containment measures are implemented from the onset.

Above all, one should avoid costly improvisations serving some ill-defined business objective that undercuts the principal projects of the network. Unfortunate examples abound among the first generation of millimeter microwave service providers, most of which spent huge sums of money leasing T1 lines to provide temporary data services to key customers and to thereby establish a presence in a targeted market. Such a ploy did nothing to position the core services of the new service provider or to give the customer any reason to adopt the new wireless services. All it did was to inflate the customer numbers while ensuring that the wireless infrastructure would never be completed.

One successful millimeter microwave network operator of my acquaintance told me that the secret of achieving profitability in wireless broadband was to build a business case for each major customer or group of customers and to calculate probable ROI on each. That way nebulous assumptions and meaningless generalizations are avoided, and the operator can make necessary adjustments in rollout plans on the fly without building up a huge unprofitable operation and then attempting to set it to rights.

Putting Strategic Planning in Perspective

Strategic planning above all must be global, encompassing every major aspect of network operations. It should also be forward thinking with the establishment of long-term growth objectives for the network and a roadmap for achieving those objectives. No service offering should be contemplated without a clear idea of how it will be supported and delivered, and no expansion plans should be made without some plausible means of acquiring future facilities. As with any other business enterprise, as little as possible should be left to chance. At the same time it should be understood that any plan regarding a communications network, especially one founded by a startup, is ultimately provisional. Too much impinging on the operation cannot be predicted, including the progress of technology, the evolution of markets, and perhaps, most of all, changing patterns of usage. In the last analysis the network operator must be adaptable and alert to the earliest signs of a change in the communications landscape.

CHAPTER 5

Strategies for Successful Deployment of Physical Infrastructures

Success in launching a broadband wireless network largely depends on three factors: creating a service model that will attract subscribers, designing the network in all its aspects around that service model, and designing and implementing a physical infrastructure that will maximize coverage and spectral efficiency. This chapter focuses on the third factor.

Selecting an Appropriate Network Topology

Chapter 4 defines the basic network topologies and examines their basic capabilities. This chapter discusses where each is appropriate.

Deploying Minority Architectures

As indicated earlier, point-to-multipoint will be used in most instances to provide last-mile access to the subscribers to the broadband service. The question then arises, When does using the other topologies becomes necessary?

Point-to-point connections are generally used in three instances: to serve a single site containing a number of high-value customers such as a business high-rise, to provide backhaul from a base station to a central office, and to serve a single high-value user requiring extremely high bandwidth such as a video postproduction house or a scientific research organization. In all three cases the full spectrum available to the network operator will generally be utilized within the single connection, and, to mitigate interference and maximize security, a very high-gain, highly directional antenna will be employed.

The tendency today is to use higher frequencies for point-to-point connections because generally abundant bandwidth is available in the higher bands and because they lend themselves to narrow-beam transmissions. The U-NII band at 5.8 gigahertz (GHz) is especially well suited to point-to-point links because it can be transmitted over long distances and because it will support throughput rates in excess of 100 megabits per second (Mbps).

Point-to-consecutive-point architectures, also known as *logical rings*, are chiefly applicable with millimeter microwave equipment. Since no manufacturer of 802.16 equipment currently makes a complete system for supporting a wireless logical ring, I will devote little attention to topology. A network operator wishing to use it will either have to jury-rig an IP-based system by utilizing Resilient Packet Ring add-drop multiplexers or use legacy SONET or synchronous digital hierarchy (SDH)-based point-to-point microwave equipment.

Mesh wireless equipment, in its current state of development, is best suited to relatively small deployments serving a few dozen subscribers. To preserve bandwidth for individual users, routes must be kept short, that is, restricted to no more than two or three hops. Hundreds of subscribers distributed over several square miles would obviously lead to longer hop sequences and slower throughput speeds, and where subscriber bases of such size must be served, the network operator would have little choice but to create a number of discrete meshes, each with its own aggregation point, or else use another topology, most probably point-to-multipoint.

Deeper into Point-to-Multipoint

Point-to-multipoint architectures, as you have seen, are the norm in pervasive broadband metro deployments and always have been. The success of such deployments depends, on the one hand, upon reaching as many potential subscribers within the area swept by a single base station and, on the other hand, on utilizing the available spectrum efficiently and effectively.

To illuminate how both objectives may be achieved, you must first examine the concept of wireless coverage areas, commonly referred to as *cells*, and how they figure in a point-to-multipoint architecture. Cells are the basic building blocks of wireless networks, and the mapping process of planned cell sites within a given territory to be covered constitutes the most basic strategic planning function of the network engineer.

Radio Cells and What They Portend

The concept of cells appears to have originated at Bell Labs, the research arm of the Bell Telephone system, back in 1948. At the time no effective means existed for automating the tuning function of individual radios to enable the concept to be realized, but later—30 years later, in fact—the Swedish telecommunications giant Ericsson would successfully demonstrate the feasibility of the concept in the first commercial cellular telephone system set up in the city of Stockholm.

Today cellular architectures are ubiquitous in wireless communications, not just in cellular telephone systems, but in wireless local area networks (WLANs), personal area communications networks such as Bluetooth, and in fixed-point broadband wireless networks of the sort constituting the subject of this book.

The Function of Cells in the Point-to-Multipoint Network

To grasp fully the cellular concept, one must first understand how traditional radio networks operate, for the cell is a fairly radical departure from older practices and, at the same time, can really only be understood in their context.

For as long as radios have been used, and they have been used for more than 100 years now, the typical approaches in two-way network communications have been either peer-to-peer linkages where one radio transmits to another, as in a citizens band radio network, or arrangements where all of the individual user terminals go back to one central base station. The latter approach typifies the wireless telephone systems predating the cellular networks, the two-way dispatch radios used in commercial fleets, and police and public safety radios.

It also typified the first attempts to deploy commercial broadband data networks in the United States.

The limitations of simple peer-to-peer, citizens band radio being a prime example, as opposed to the similar but much more sophisticated mesh, are fairly obvious. Each individual radio can reach adjacent radios only if it is not to broadcast interference far and wide, so such an architecture cannot form the basis of a high-speed access network.

The limitations of the second approach are not so obvious, but they are nonetheless real. Simply stated, a single base station has only a certain amount of spectrum to utilize to reach all the subscribers in a given geographical area. Thus, in early mobile telephone systems where at most a few dozen channels were available, a few hundred subscribers were all the network could support using a widely accepted ratio of ten subscribers to every one of the available channels. (Such a ratio is based on the concept of statistical multiplexing, where statistically there may be only a 10 percent chance of any one subscriber transmitting at any given instant.)

Obviously, a mass market service such as cellular telephony must support not hundreds, but thousands or even tens of thousands of subscribers within a metropolitan area, so it demands something more, either a vast amount of spectrum or some way of reusing a more limited number of channels so that more than one user can occupy a given channel simultaneously. Given the total demand for spectrum on the part of a huge array of powerful interests, the vast amount option is not really available; at least it was not until the emergence of ultra-wideband radio recently. The only choice the operator had was to find some means of reusing spectrum, and the cellular approach, described next, spoke to that need.

So here, in brief, is how cells figure in a radio network and permit extensive frequency reuse: Radio cells themselves are relatively small areas surrounding base stations. Both the base station and the subscriber terminals within the cells transmit at very low power so that the signal quickly fades out to the point where it will not interfere with someone else on the same channel a fairly short distance away. Thus, each cell becomes in effect a subnetwork.

In principle this is simple enough, but, for the scheme to succeed, one has to work with quite high frequencies that will fade quickly over distance, and during the first two decades following Bell Labs' enunciation of the principal of cellular reuse, the only devices that could transmit at such frequencies were exotic vacuum tubes such as klystrons and traveling wave tubes, neither of which was suitable for inclusion in a subscriber terminal. Moreover, if one wanted to use cells in a mobile network, further problems presented themselves because a terminal in motion would always be passing out of range of particular base station and had to have some means of instantly acquiring an unoccupied channel in an adjacent cell if a transmission was to be maintained. Such "handoffs," as they have come to be known in the cellular industry, require the use of powerful computers and specialized software at the base station and some computing ability in the terminal itself. The microprocessors that could support such functionality in a handset were not available much before the opening of the first network in 1978.

Since cellular telephones established themselves in the early 1980s, the principle of cellular reuse has also been applied to WLANs and internal wireless phone systems called wireless PBXs and of course to fixed broadband wireless metro networks. In the case of the latter, handoffs are not ordinarily required, so the design of the network is somewhat simplified over that of a mobile network, but the basic design principles are much the same, as are the benefits, chief among them reduction of transmit power requirements for terminals and vastly increased spectral efficiency over the entire network because of aggressive frequency reuse.

I point out, however, that the cells deployed within broadband wireless networks are usually larger than those in cellular telephone networks.

A Note on Channel Assignments from One Cell to Another Concerning this matter of frequency reuse, it should be understood that channels, according to standard engineering practice, cannot be reused from cell to cell because a signal fades only gradually over distance and will not be sufficiently attenuated so as not to cause grave interference if another user tries to occupy the same channel in an adjacent cell. Therefore, a channel would normally be reused only one cell diameter away at best. Figure 5-1 shows frequency reuse patterns in a cellular network; for simplicity's sake, it displays only four channels. The exception would occur when the cell was divided into sectors, which are discussed next.

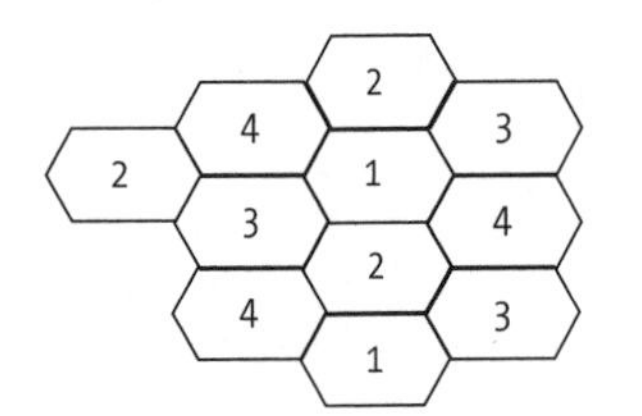

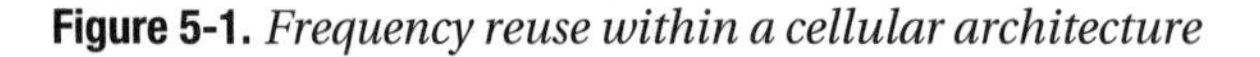

Figure 5-1. *Frequency reuse within a cellular architecture*

Today this standard engineering practice has been subject to some modification because of the appearance of advanced modulation techniques such as direct sequence, Code-Division Multiple Access (CDMA), and orthogonal frequency division modulation (OFDM); the introduction of polarization diversity; and the emergence of smart antenna technology. All these techniques, described in detail later in this chapter, increase the immunity of a transmission from interference and allow reuse of a channel at reduced distances—in other words, less than one cell diameter away. Reuse, at least within certain sectors in adjacent cells, does then in fact become possible, though certain minimum distances still have to be maintained. Coincidentally, simple rules of thumb for channel spacing become increasingly difficult to formulate.

As it happens, numerous schemes and formulae have been developed for cellular telephone networks for optimizing frequency reuse patterns, but all of these, understandably, are optimized for mobility and must take into account the fact that a subscriber terminal must be able to "see" two base stations simultaneously at the boundary of a cell in order to initiate a handoff. As a result, such techniques cannot be imported into the broadband fixed wireless arena without extensive modification. Considerably less attention has been given to maximizing frequency reuse in fixed broadband networks through improved calculations for base station placement or through new protocols for dynamic channel assignment.

Part of the problem has to do with the ad hoc nature of most fixed wireless broadband networks to date. Typically, wireless broadband networks have been the product of startup companies with limited resources that have tended to add capacity as needed with no overall plan of what a fully loaded network would look like and with no staff with either the training, inclination, or mandate to refine the formulae used by cellular network engineers to perform cell mapping and base station distribution. The tendency instead has been to rely on relatively crude cell-splitting techniques to accommodate increased traffic and, ironically, on ultra-sophisticated adaptive modulation and beam steering technologies for the same purpose, neither of which, incidentally, require much engineering skill on the part of the network operator to execute. The first can be accomplished simply by building new base stations and backing down on the power levels of those already in operation, and the second relies on the intelligence programmed into the radio itself and not on the abilities of the network engineer.

However, a few specialized software tools exist for frequency reuse planning that have recently become available for broadband wireless operators, and that you will consider in the following section devoted to explaining the basic principles behind cell mapping.

Overall Strategies for Cell Mapping

The first requirement of any strategy for cell mapping within a broadband network is flexibility and the scope to be able to factor in a vast range of transmission speeds and data rates as well as highly differentiated service offerings and user profiles. Obviously this entails considerable complexity in the planning tools.

The older cell mapping procedures utilized by mobile telephone operators are themselves quite complex, but in many ways the situation confronting the fixed broadband wireless operator is such as to demand even more involved planning and plotting methods. Cellular telephone networks generally deliver fairly uniform service offerings, whereas wireless broadband networks must accommodate varying throughputs, some of which may be instantaneously user defined; a multitude of different service levels; and connections having widely varying fade margins. Furthermore, fixed broadband networks often have different subscriber densities and traffic volumes within various areas, but usage patterns are apt to be relatively uniform in mobile networks.

The very large number of variables make network planning difficult in broadband fixed-point networks, but both rules and resources are available, and the network operators should make themselves aware of them before a single base station is installed.

Network Mapping Software Tools

The network operator should always strive to plan the network systematically utilizing available software resources. Two companies in particular, Florida-based RadioSoft and Oregon-based EDX, specialize in offering mapping software to wireless operators. Both enjoy good reputations in the industry, and both support their respective products with consultation. True, both software packages incorporate known engineering formulas that can be abstracted from standard network design texts, but assembling such information and then attempting to apply it to the latest radios with the most abstruse advanced modulation schemes is an engineering exercise that could consume weeks. It is better to utilize the existing automated tools and improvise only when absolutely necessary.

I particularly recommend the EDX software because it is so comprehensive. EDX Signal-Pro can factor in distance, blockage, radio sensitivity and selectivity, the use of advanced modulation such as OFDM and W-CDMA, adaptive modulation, a number of smart antenna technologies, and antenna polarization. In addition, it can set channel reuse patterns based on whether channels are defined as frequency divisions only (frequency division duplexing, or FDD) or as time slots within frequency bands (time division duplexing, or TDD). It can also profile every uplink and downlink as well as provide global readings for interference within an entire sector. Furthermore, it can calculate network capacities with various mixes of traffic—such as best-effort data, voice, multimedia, and so on—and includes three-dimensional mapping software, which shows building elevations and transmission paths. EDX software is widely used by the larger wireless carriers, but because of the considerable cost of a complete package—upwards of $50,000—the smaller carriers that probably need the software the most have been reluctant to use it. One solution for the small operator who cannot bear the expense of purchasing the software suite initially is to utilize the services of a consultant who owns the package. Figure 5-2 shows the use of EDX software in Chicago.

Figure 5-2. *EDX software, courtesy of EDX*

Bear in mind that infrastructure products differ in their capabilities and that some of the newer technologies such as multiple-in, multiple-out (MIMO) antennas and frequency-agile radios may not be completely characterized in existing planning tools. Therefore, if one is contemplating the purchase of innovative technology to increase throughput or frequency reuse, one must make certain the vendor is willing to take an active part in the network planning process.

Network Mapping Is Always Provisional

Cell mapping is highly specific in terms of frequency band selected; modulation technique; the sensitivity of the radio; the types of services supported, such as best-effort high speed, access, voice telephony, rich multimedia, full mobility, and so on; and the distribution of subscriber sites. Should the network operator decide to transmit at another frequency than that used initially or attempt to offer services requiring constant throughput rates, the specific network architecture in use previously may not be adequate. It follows then that if the network at first offers only basic access but the intention is to move toward diversified services, then the network should be overdesigned in terms of basic access. A network that can support advanced multimedia well can nearly always provide basic access, but the converse is not true.

The Importance of Projecting Growth Patterns

A third consideration is closely related to the second. One should always keep in mind that bandwidth can be throttled down from a set maximum but never throttled up, at least not without major equipment upgrades. It follows that each cell should be designed to deliver however much capacity is likely to be demanded by the customers over the lifetime of the network. If, for instance, a large business park full of 20 likely high-usage customers is situated two miles from the nearest base station in a 5.8GHz network, one may want to consider building a base station nearer to those businesses or at least securing location rights that will enable one to do so in the future. One may be able to serve the single subscriber that one currently has in that business park perfectly adequately from two miles away, but one may not be able to serve five or ten. In other words, the mapping of the network must extend into the future and must anticipate the eventual maturity of the network with the full complement of base stations that will eventually have to be in place. Such projections will never be entirely accurate because subscriber take rates, equipment advances, and changes in urban topography are never completely predictable. One can make informed guesses as to growth and change in the network, however, and then model a distribution of base stations based on sound engineering principles.

A final general design principle is to provide oneself with choices in terms of base station placement. In planning for the future one can never be absolutely certain that a desired site will be available when one is ready to occupy it. A building where the owner has agreed to provide roof rights may be sold. An antenna tower that has space today may be filled tomorrow. If one does not have options at the time when expansion is indicated, the network may not be able to reach potential customers.

Macrocells and Their Limitations

Because a cell is the basic constituent unit of a point-to-multipoint network, the distribution of cells determines the coverage and capacity of the network and of course has a major bearing on capital costs. Thus, cell size matters very much.

Network engineers tend to think of radio cells in terms of the areas they encompass. A large cell with a radius of miles is known as a *macrocell,* while cells of a kilometer or two in radius constitute *microcells.* Cells with radii measured in the hundreds of meters are *picocells.*

The usual pattern of broadband wireless operators, as indicated in Chapter 2, has been to launch the network with a single base station defining a single cell and with sufficient transmitting power to reach all or most of the potential customers within the metropolitan market addressed by the network operator. Both low-frequency microwave and millimeter microwave operators have used this type of single-cell architecture.

Lower-frequency microwave operators have been able build macrocells with radii exceeding 15 miles, and if they utilize sectoral antennas—actually clusters of high-gain antennas distributed around a central mounting pole—they can reuse the available spectrum as much as four times over in theory, eight being the maximum number of sectors that is practical with conventional nonadaptive antennas and four being the maximum reuse factor within a single cell. It would appear to follow, then, that if one starts with a generous frequency allocation—say, the nearly 200 megahertz (MHz) available in the Multichannel Multipoint Distribution Service (MMDS) bands—and multiplies that by four and assumes a three-bit-per-hertz ratio, then one is looking at minimally a couple of gigabits per second total system throughput, a rate that translates into quite a respectable system capacity.

None of these theoretical maxima can be realized in practice, however, and they cannot even be approached. The narrow beams that define each sector will spread out over distance and begin to interfere with one another at the outer edge of the macrocell so that all eight sectors cannot be extended the full width of the cell. Then too, bit rates drop off with distance as the signal attenuates, the fade margin declines, and the error rate ascends. In fact, real throughput rates are apt to be half or less the theoretical maxima at the farthest transmission distances.

In practice, only about a gigabit of capacity is likely to be reliably available, even assuming that generous 200MHz spectrum allocation. If you settle on an oversubscription rate of four to one, a fairly conservative figure, then the network can accommodate 400 subscribers at a 10Mbps throughput rate for each albeit with no service guarantees. If one can charge appropriately—say, $500 to $1,000 per month—that may be the basis of a sustainable business, at least in the short run, though an annual revenue in the $2 million to $4 million range may not permit the network operator to capitalize the expansion of the network.

Furthermore, in such a macrocellular architecture—absent the newer NLOS technologies—obstructions will put many potential subscribers entirely out of reach. What percentage of potential subscribers cannot be served is a matter of dispute among authorities, and in fact the particular topography in which the network is being deployed can give rise to enormous variations in this regard. By most accounts, at least 40 percent—and perhaps as many as 80 percent—of potential subscribers are unlikely to be served because of the presence of obstructions.

Faced with such constraints, network operators can resort to a number of ploys. They can use repeaters to reach certain subscribers, a sort of partial cell-splitting approach. They can employ sectoral antennas to achieve frequency use within the cell. They can employ dual polarization to reuse spectrum. They can resort to full-on cell splitting and begin to create a microcellular network. Or they can use NLOS technologies to access out-of-reach subscribers, though only in the immediate vicinity of the central base station and not at the outer periphery of the macrocell. They can also do all of these. But what they must do in all cases is to increase spectral efficiency, that is, overall carrying capacity of the networks based on available spectral resources.

Principles of Frequency Reuse and the Technologies for Achieving It

In every wireless network with multiple users and limited spectrum, the network operator is forced to confront two conflicting demands: the need for aggressive frequency reuse and the need to mitigate interference. Put another way, a channel can only be reused when interference is reduced, but reuse of a channel inevitably increases interference levels! Frequency reuse conduces to an expanded customer base and maximum exploitation of available spectrum, but unless the mitigation of interference can be achieved in some measure as well, the quality of service available to the subscriber will be unacceptable. The pleasant irony here is that the technologies that improve frequency reuse generally reduce the effects of interference as well, but trade-offs are always involved; in other words, very pronounced frequency reuse will increase interference, and minimizing interference will limit the degree of frequency reuse possible.

Use of Repeaters

A *repeater* consists of an antenna and a simple radio transceiver without much intelligence behind it. A repeater normally uses a simple point-to-point connection back to a remote base station, which in turn communicates with a central office base station. A repeater has no switching or routing capability; it simply extends the reach of an individual base station in a given direction, and is often used to reach a few isolated customers whose numbers do not justify the creation of a complete new base station.

The installation of repeaters is a tactical move on the part of the network operator, one that is indicated only in certain circumstances. As a rule, a repeater is less expensive than a full-fledged base station because of the lack of switching and/or routing equipment, though one still has to pay for the site and the radio. Because of its lack of intelligence, the repeater cannot really augment the capacity of a network appreciably, but what it can do is enable the operator to extend the boundary of the cell in one direction in order to encompass a few subscribers who would otherwise be unreachable.

Sectorization

I have already touched upon sectorizing slightly. In this section I will attempt a more complete explanation.

Sectorizing is using an array of highly directional antennas to direct intense radio frequency (RF) energy into a designated area of the cell and little energy elsewhere. The sector defined by the antenna array appears like a pie slice when depicted on a diagram. Sectorization itself is a species of *spatial diversity,* of which adaptive beam steering is another. In both cases the operator is able to define subchannels in three-dimensional space rather than by frequency division or the use of sequential time slots. Figure 5-3 shows how a directional antenna defines a sector by sweeping a narrow arc. The figure shows a directivity polar plot for such an antenna.

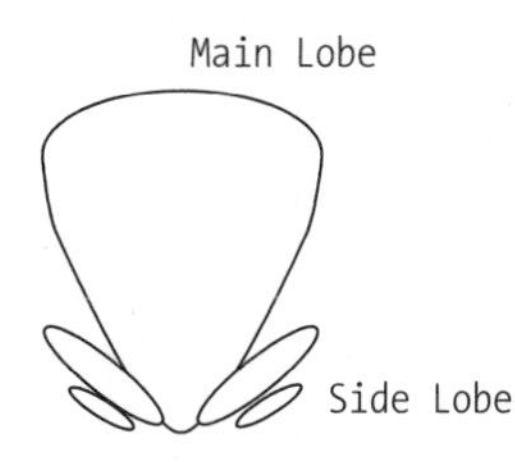

Figure 5-3. *Directional antenna directivity polar plot*

Sectorizing is somewhat akin to cell splitting and indeed may be viewed as a sort of poor man's cell splitting inasmuch as it allows the operator to reuse spectrum aggressively without installing a new base station and defining a new cell. It is a standard tactic in the majority of broadband wireless networks.

Sectoral antenna arrays vary according to the number of sectors they form, with three to eight sectors being the usual range and four and six being the most common numbers. Obviously, the more sectors, the narrower the beam width of each antenna in the array. Some arrays are configurable to cover the whole range from three to eight, that is, the number of sectors that can be made to vary by adjustments in the antenna itself. Since each antenna in a sectoral array will be provided with a separate radio, sectorization definitely entails higher costs for the network operator.

Sectors are akin to cells in that they ordinarily permit a channel to be reused one sector away but not in intervening sectors. This means that when four sectors are present, the reuse factor is two, and for six the factor is three, and so on. As indicated earlier, advanced modulation techniques loosen such reuse constraints to some extent.

In very large cells, as you have seen, the narrow beams formed by sectoral antennas spread out over distance, so actual frequency reuse capabilities will be reduced at the outer periphery. Naturally the bigger the cell, the greater the spreading and the more the sectors will overlap.

Sectoral antennas can be used in both the lower microwave and millimeter microwave regions, though the physical form of the antenna will differ with frequency, with horns and waveguides being employed at the highest frequencies and arrays of spaced omnidirectional pole antennas at the lower frequencies. Sectoral antennas of whatever type are considerably more expensive than the omnis used in WLAN applications, but, given the vastly increased spectral efficiency that they confer upon the network, the cost is trifling. Indeed, using such devices has almost no downside, though they do concentrate the radiated energy, somewhat increasing the potential for interference outside the cell. The cure for that is to polarize the antennas in the horizontal plane so that only horizontally disposed magnetic fields are propagated. The whole array is then tilted downward so that beyond a certain distance the radiation will simply be absorbed into the ground. The following section explains polarization itself.

Polarization Diversity

The term *polarization* refers to a property common to all radio waves, namely that the magnetic waves emanating from the antenna tend to propagate outward in a shallow ellipse. If the antenna element is upright, the wave propagation will be in the horizontal plane, and if the antenna itself is horizontal, a vertical propagation pattern will occur. In either case the wave front is said to be *polarized* in one or another dimension—or *linearly polarized*, to use the technical term. The third dimension, through which the airlinks extend, is occupied by wave fronts in either state of polarization as they make their way toward the receiver site.

Both transmit and receive antennas are generally polarized in the same dimension, and if they encounter signals of the opposite polarization, that is, offset by 90 percent, they will interact with relatively few magnetic lines of force and will not develop signals of much strength. The result is that signals of opposing polarization will not interfere with one another even if they occupy the same channel. Signals of opposite polarization, incidentally, are said to be *orthogonal* to one another.

In self-installs, particularly those involving indoor antennas, polarization is apt to be haphazard, which argues against the use of an airlink based on simple linear polarization. One solution is circular polarization, described next.

Now it is also possible to tilt antenna elements at intermediate angles and thus create a multitude of polarization states, but in such cases, the various states will not be orthogonal to one another and will interfere. Nevertheless, radios have been created that could resolve several nonorthogonal polarization states and reuse spectrum very aggressively in this manner. At this time, however, no such radio is available for broadband operators. What is available are radios that offset two signals 45 degrees from the vertical so that both antenna elements are tilted. The total separation is still 90 degrees and thus fully orthogonal, but the propagation patterns tend to be more useful, though dual vertical and horizontal polarizations are employed as well in broadband networks, such an arrangement being the aforementioned circular polarization. Circular polarization is usually provided to a single radio, and its purpose, as indicated previously, is to afford the best reception in a random polarization environment. Figure 5-4 shows the various polarization states.

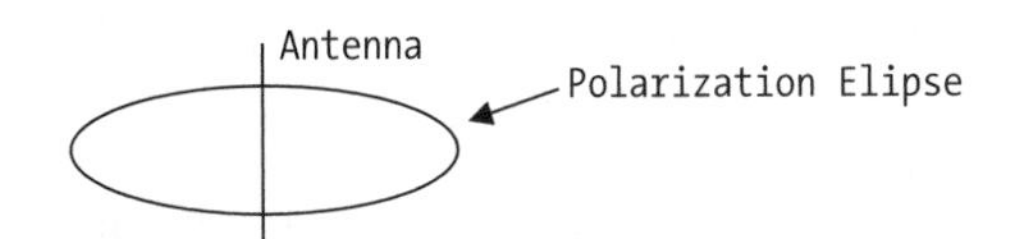

Linear Polarization in the Horizontal Plane

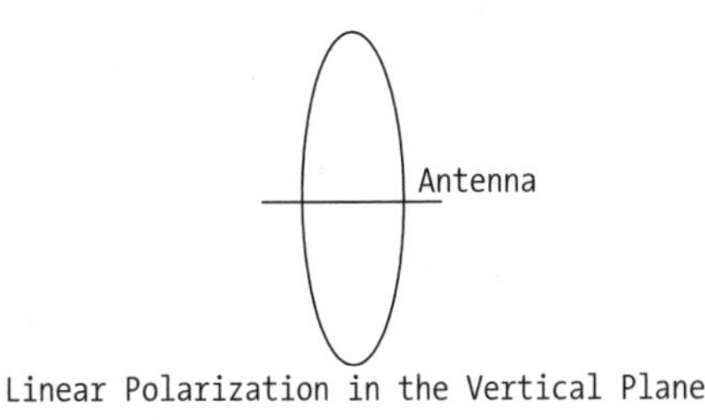

Figure 5-4. *Polarization states*

The property of polarization can be exploited to reuse spectrum within the same space by setting up airlinks of opposing polarity, a tactic known as *polarization diversity*. Reuse, however, will reach only a factor of two by this means, so polarization must rank among the weaker methods for enhancing spectrum reuse.

Dual polarization can be used in tandem with spatial diversity via sectorization or adaptive arrays, but I know of no commercial product with such capability, though an experimental system developed by Lucent was so enabled. In most systems one or another strategy is adopted, that is, polarization diversity or spatial diversity.

In sum, polarization diversity is part of the network operator's bag of tricks for extracting the best performance from a particular radio in a particular RF environment. The aforementioned EDX software has subprograms for plotting the effects of polarization diversity on reception.

Cell Splitting

Cell splitting consists of decreasing the radii of existing cells and adding new ones. Cell splitting has been one of the principal means by which cellular telephone operators increased the capacity of their networks, and it will also be a standard tactic for broadband wireless operators, although it will be supplemented by NLOS technologies that were not available to cellular operators during the period of greatest growth in cellular networks.

Cell splitting should properly be considered a species of cell mapping or planning and refers to a process by which the network operator redetermines the minimum number of cells required to provide the desired coverage and capacity as the network attracts more subscribers. One does not just split a cell into two neat halves; one has to construct entirely new coverage patterns for each of the resulting base stations.

It is a fairly involved process because of the very indefiniteness of cells themselves, a fact that may not be immediately apparent to the individual without extensive knowledge of RF propagation.

When a diagram is made of a cellular network—and I use the term broadly here, not just with respect to mobile telephone networks—the cells are often represented as a sort of honeycomb pattern, a hexagonal arrangement of spaces where everything is clearly defined and fits neatly together, though occasionally a checkerboard pattern is substituted. Both patterns are abstractions, and misleading ones at that.

A cell radius is always an arbitrary value. A radio signal does not abruptly cease to propagate at so many yards from the transmitter; indeed it continues to the very edge of the cosmos, though growing steadily weaker over distance. What this means is that not only do adjacent cells overlap, but that every cell in the network overlaps with every other cell. Picture the propagation of signals as ripples or wavelets spreading over the surface of a pond. Each pattern of ripples travels everywhere, and each reflection begets new ripples. This being the case, cells should be considered as concentrations of RF energy rather than as well-defined geographical areas.

Nevertheless, subscriber units within a point-to-multipoint network architecture must treat the cells as if they were well-defined entities; that is, subscriber units located at the arbitrary boundary separating cells must communicate with only one base station even though they are receiving signals from several, albeit at reduced levels. In other words, they must lock onto the base station with which they are registered and reject interference from all others, and for that to happen transmit power levels must be strictly controlled throughout the network.

This, as it happens, has important implications for cell splitting. Because the average radii of all the cells in the network decrease with cell splitting, so perforce does the transmit distance from the subscriber terminal to the base station, and vice versa. And because of the shorter distances involved, transmitting power must be reduced at both the base station and the subscriber terminals to avoid interference throughout the network.

Because in a fixed broadband network (excepting the mesh variety, which really does not have cells as such) subscriber units are normally assigned to a specific cell (an assignment that is ultimately based on the strength of the signal in either direction), redetermining the optimal transmit power levels becomes extremely important during cell splitting.

Cell splitting may also necessitate the reassignment of channels within each cell in the network since the new cells will be establishing new channel relationships with surrounding cells. Altogether, it is not a process to be undertaken lightly and without a thorough reexamination of the entire network.

As indicated earlier, it is advisable to plot out the location and capacity of every base station that the network will ever need at the time the network is being launched, though that may not always be possible, and the time may come when the network operator is forced to consider unanticipated microcells to meet demand. At that point, the network operator is faced with the task of essentially reengineering and rearchitecting the entire network. Obviously, there are limits to what can be done here. For practical reasons one is not going to relocate existing base stations. But power levels and channel assignments will all have to be redone, and the same software tools used in the initial planning process will have to be used all over again.

Line of Sight and Non–Line of Sight

I have previously alluded to NLOS many times in this book. In this section I present a fuller explanation of how it is achieved and what it means in terms of network mapping.

First a bit of background: Radios operating in the region above 2GHz, the region that has generally been assigned to broadband wireless services, are, as you have seen, easily obstructed. This is a simple matter of the physics of wave propagation at such frequencies. Since the degree to which such high-frequency transmissions are obstructed has a major bearing on the broadband wireless operator's ability to sign up customers, the equipment segment of the industry has long sought some means of getting around or through obstructions. The collective term for the technologies for doing so is non–line of sight (NLOS).

NLOS is a term used rather loosely in the industry—too loosely in my opinion. The term applies to a number of distinct technologies of varying capabilities and maturity, and the performance of so-called NLOS equipment in the field is far from uniform. While manufacturers that have appropriated the NLOS nomenclature are fond of claiming that their respective products will double the effective coverage of a network, and in some cases can frame fairly plausible arguments based on enhanced link budgets, such claims must always rest upon certain assumptions regarding the distribution of obstructions within the locale in question, assumptions that may not necessarily hold true in individual instances. Thus, no claim should be taken at face value, and any radio that has to be used in a NLOS setting should be thoroughly tested in the environment in which it is to be used before it is purchased.

NLOS, as I will use the term, refers to any technique for lessening the effects of physical obstructions, and the emphasis is on the word *lessening*, because no NLOS technique can entirely eliminate the effects of blockage. The success of the technique is directly measurable in terms of the signal strength at the radio front end, the block of circuitry where actual demodulation of the signal begins. NLOS is never an all-or-nothing proposition. It is a matter of greater or lesser signal strength—it is that simple.

Most of the manufacturers that have survived in the public broadband wireless arena now claim to manufacture NLOS equipment. Within the industry overall, a near consensus has been reached on the matter, and that consensus is that networks of any size operating in the lower microwave regions are not viable without NLOS equipment. I am inclined to accept this consensus on purely empirical grounds, leaving aside precise definitions of just what NLOS is, because, simply stated, the attrition rate among network operators attempting to use strict line-of-sight first-generation equipment has been horrendous. Obviously, other factors are at work as well, but the inability of operators to reach customers without line of sight to a base station limits absolutely the size of the subscriber market. And that this limitation may be such as to rule out half of the potential customers in a network of one kilometer radius cells does much to explain the failure of so many of the pioneers.

What NLOS Means in Terms of Wave Propagation

The object of this book is to go light on RF theory and heavy on practicality. However, to understand even the rudiments of NLOS, one must know something about how physical objects affect radio wave propagation.

When a radio wave encounters a physical object—and by that I mean anything from an air molecule to a mountain—it can behave in one of three ways. It can give up some of its energy to the object in the form of heat, a process known as *absorption*. It can rebound from the object without surrendering appreciable energy to it, a process called *reflection*. Or it can bend around the object, a process known as *diffraction*. These three processes, by the way, are not mutually exclusive. A reflected signal, for instance, may immediately be diffracted as it encounters a different contour of the object reflecting it, and in every case where reflection or diffraction occurs, some energy will be absorbed as well.

Absorption

Pure absorption does not change the direction of the radio wave but robs it of energy and reduces the fade margin. Since the signal steadily loses energy simply by being propagated over free space, ultimately limiting its useful range, the effect of absorption of energy by physical structures such as walls or trees is to further reduce the distance over which a reliable connection may be maintained, and often very appreciably at that. Indeed, absorption losses can, if significantly severe, interrupt the signal entirely. A good example of this is provided by a large park full of high trees where anyone attempting to blast through the treetops with a microwave signal to reach buildings on the other side will be completely unable to establish an airlink even with the most advanced NLOS equipment. Incidentally, as a good rule of thumb, a direct signal path through dense foliage will result in a loss of signal strength of approximately 1 decibel per meter.

Reflection

The effect of reflection depends heavily on whether the main lobe or the side lobes are reflected. If only the side lobes are reflected, multipath distortion will occur in the main lobe, compromising signal integrity and resulting in loss of data but not interrupting the signal entirely. If, on the other hand, the main lobe is reflected by an obstruction standing directly in its path, then almost no energy from that lobe will appear at the receiver's antenna. In such instances, reflected energy, most likely from a side lobe, may reach the receiver at a sufficient level to provide a usable signal, although the fade margin will be vastly reduced over what it would be with a direct signal. The problem here, however, is more significant than a simple reduction in signal level because now the receiver is operating entirely in the multipath environment, and it is likely to be subject to not one but multiple reflections, each of which will interfere severely with one another, further reduce fade margin, and substantially increase the bit error rate.

To get a better idea of how multipath occurs, refer to Figure 5-5 for a schematized depiction of multipath reflections. The severity of mulitpath will depend on the frequency of the transmission, the distribution of reflective surfaces in the area separating the transmitter from the receiver, and the directivity characteristics of the transmitter receiver.

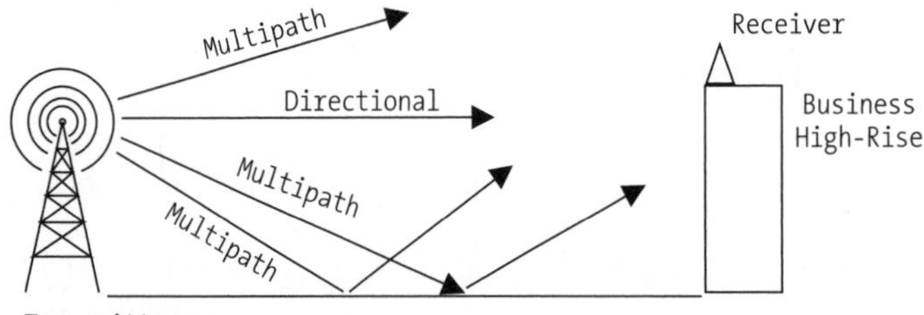

Figure 5-5. *Multipath reflections*

Now it is entirely possible to set up a deliberate reflection in an effort to reach an obstructed receiver, but again one is likely to be operating in a multipath environment because the side lobes will be encountering their own obstructions, possibly the same obstruction affecting the main lobe, and will likely be reflected into the path of the main lobe. In addition, the direct signal will lose energy to whatever object is reflecting it.

A number of radios on the market do have the ability to function in a pure multipath environment, but the important thing to remember is that they do not function nearly as well in such a regime. Thus, the presence of some degree of NLOS capability in a receiver does not mean that one can ignore line-of-sight considerations.

Diffraction

Diffraction, which occurs when radio waves bend around the edges of an object, results in a transmitted beam becoming off-axis in relationship to the receiver antenna. Since radio waves are vastly greater in wavelength than visible light waves, diffraction can occur when there is still optical line of sight between two radio terminals. The effect of diffraction is to reduce signal strength substantially at the receiver and also to introduce a possible source of unwanted early reflections. Here again, a diffracted signal is not necessarily useless, but it is certainly less useful than a direct signal.

NLOS: A Truer Conception

It should be obvious by now that, absent line of sight, no radio functions optimally, and the mere fact that a radio has some measure of NLOS capability does not mean that the network operator can place base stations and subscriber haphazardly. NLOS technologies are partial remedies (*misleading* might be a better word) at best. They do not rewrite the laws of physics.

Before I leave this discussion, I should mention the term *near-NLOS*, industry jargon that surpasses simple NLOS in its sheer ambiguity. As the term is most commonly employed, it refers to radio equipment capable of dealing with consequences of obstructions that occur within the Fresnel zone but do not block optical line of sight. What is really being claimed here is not the ability to reach completely obstructed sites but the ability to cope with multipath with a high degree of effectiveness.

NLOS Technologies

Radios are able to function in NLOS environments by using a number of technologies. These involve the basic functioning of the radio, the modulation system employed, the design of the antenna, and the use of certain extraneous signal processing techniques normally involving multiple antennas.

Basic Radio Performance and NLOS

Increasing radio sensitivity and channel selectivity while providing the radio front end with high overload capabilities and a very low noise circuitry will in and of itself endow the receiver with an enhanced ability to use the weakened signals upon which one is forced to rely on a NLOS installation. In other words, basic high quality RF engineering and circuit implementation, most of which, incidentally, is still analog, will go a long way toward the aiding in the recovery of a marginal signal.

Modulation Technique and NLOS

Modulation technique can also play a major role in enabling links where line of sight is not present. All of the spread spectrum techniques in common use in broadband wireless, including frequency hopping, direct sequence, CDMA, and OFDM, can immunize the signal to some extent from multipath, with frequency hopping perhaps performing best in this regard. However, interestingly frequency hopping figures in the IEEE standard only for WLANs and 802.11, not in the 802.16 standard. OFDM also has very good immunity from multipath, and OFDM radios can generally operate in the presence of weaker signals than is the case with most other modulation techniques, thus permitting the radio to retrieve a usable signal in marginal settings. Nevertheless, OFDM, in and of itself, will not allow a link to be maintained in the face of complete line-of-sight obstructions, as is the case when a target terminal is located behind a hill, a large masonry building, or a dense grove of trees. It is actually better suited to near-NLOS conditions.

OFDM is the only advanced modulation technique specified in the 802.16 standard, and only in the 802.16a amendment. Single-carrier modulation is permitted within the 802.16a standard as well, but no company currently active in wireless broadband within the lower microwave region is making a single-carrier system any longer except for wireless bridges.

OFDM

Since OFDM figures so prominently in the standard, and because its advocates make so many claims in its behalf, the network operator should acquire at least a rudimentary understanding of how the technology works, of its real strengths and weaknesses (and it has weaknesses), and the degree to which it truly advances the art.

OFDM has a lengthy patent history and a long incubation period, but in its present form it is quite a recent development. Indeed, prior to 1990 the basic technology that would permit the concept to be realized scarcely existed.

In simplest terms, OFDM is a technique by which a message is assigned to a number of narrowband subcarriers, usually numbering in the hundreds or thousands, simultaneously. In 802.16a the specified number of subcarriers is 256. The same information is not replicated on all of the subcarriers; rather, it is parceled out, with some bits going to one subcarrier and some to another and to another. Some redundancy will normally be present; that is, some bits will be shared among some sets of subcarriers, but, in the main, the data will be widely scattered over a considerable expanse of spectrum.

Incidentally, OFDM may not be considered a modulation technique in the strictest sense because there is nothing in the technology itself that determines how a signal is impressed on a carrier wave, and OFDM can be—and in fact always is—combined with amplitude modulation, phase modulation, or combinations thereof. It can also be combined with direct sequence or CDMA, both of which involve an initial modulation of phase to impress the signal on a carrier and then remodulation of the resulting signal to impose an additional coding sequence upon it. OFMD is in fact combined with frequency hopping in Flarion Technologies' Flash OFDM system.

■ **Note** For those who lack a technical background in radio engineering, a carrier is a wave of fixed frequency that is made to vary in some way; the variations constitute the encoding of information on the carrier.

In essence OFDM represents a fairly simple idea, and the basic concept goes back to the 1950s, but when attempted with 1950s RF technology, it did not provide a lot of benefits except to military radio personnel attempting to render transmissions difficult to intercept. What made OFDM both feasible and intriguing was the development of cost-effective digital signal processors capable of performing digital fast Fourier transforms (FFTs), covered shortly, on very high frequency waveforms, and such digital signal processors (DSPs) did not exist until as recently as a few years ago.

To explain the significance of digital FFTs to OFDM, you need first to look at the design goals in engineering an OFDM radio, as follows. First, one wants to pack available spectrum very efficiently; second, one wants to achieve a very high immunity to interference and multipath; and third, one wants to reuse spectrum aggressively. OFDM does all of these things thanks to a DSP programmed to perform the FFT.

A *Fourier transform*, for those lacking a grounding in the mathematics of wave theory, is a computational method for deriving the single-frequency constituents of a complex waveform, and it enables a radio to tune to various frequencies entirely in the digital domain since a microprocessor is actually running the computations. This is in marked contrast to the operation of a traditional analog radio where electrical filters separate individual frequencies and no computing functions are performed.

A Fourier transform requires a powerful processor, which adds considerable expense to the system, but it can disentangle transmissions that are so close together in frequency as to overlap, and it eliminates the expedient of empty guard bands between channels that are required in all analog radio systems (most so-called digital radios used in cell phones and WLANs still use conventional analog front-end tuning circuitry).

The end result is that the FFT-based OFDM system can pack the available spectrum with a great multitude of overlapping channels and tease them out of a blizzard of interference by finding the repetitive patterns that signify a carrier wave while simply ignoring extraneous information. Because the resulting system is at once so robust and so spectrally efficient, it lets the network operator reuse subcarriers very aggressively and tolerates degrees of multipath that would cripple most other systems. It also allows the receiver to recover a weaker signal than is possible with most other modulation techniques and thus to operate in the presence of obstructions that would render other radios useless.

The fact that the bit rate for individual subcarriers is relatively low also contributes to the system's high immunity from multipath. Because individual bits endure for relatively long durations, antiphase reflections are less likely to result in complete cancellations of an individual bit.

The drawbacks? The main problem with OFDM is electrical inefficiency. The DSP gobbles current, and the subcarriers are always transmitting regardless of whether any data is assigned to them. For this reason, OFDM has never been used in a commercial mobile system to date and is not favored by most of the proponents of 802.20. Nevertheless, OFDM is favored by the architects of the fourth-generation cellular telephone system and may see deployment there by the end of this decade.

Another limitation with OFDM is that the technology is not really applicable above a few gigahertz. Existing logic circuits simply cannot switch fast enough to deal with signals in the millimeter microwave region, and even if they could, it is not clear what OFDM would buy the operator there since multipath is not a problem, and signal attenuation through the lightest obstructions is so severe that no modulation technique is going to be much help in preventing it.

Nevertheless, OFDM will probably gain ever-wider acceptance in the future and may even find its way into mobile networks. The distant future is difficult to predict, though. OFDM faces challenges, at least within the 802.20 standard, from the older technology wideband CDMA, and may face challenges in all broadband wireless applications from the rapidly evolving ultrawideband RF technology. What I can say with utter certainty is that OFDM does provide a means of building useful NLOS links at least in some circumstances, and when combined with other technologies, such as adaptive array antennas (covered next), can enable high degrees of synergy.

Multiple Antennas: Diversity Antennas, Phased Arrays, and Smart Antennas

For the worst-case scenarios, something more is required than advanced modulation, though even that something may not be enough in all cases. The following sections refer to multiple antennas, conjoined in most cases with a sophisticated signal processor that can sample the different antenna feeds and construct useful signals where none may be present with a single antenna element.

Diversity Antennas The simplest kind of multiantenna array is what is known as a *diversity antenna* system. Here two or more antennas, generally simple omnidirectional rods spaced some distance apart, are deployed. These work essentially by choosing between or among different samples of the same signal.

At the wavelengths used for broadband wireless services, the signal quality may differ considerably from one antenna to another. The radio behind the diversity antenna will have circuitry for detecting the signal least afflicted with multipath distortion and will select that antenna receiving such a signal. In cases of changes over time in the incidence of multipath from one antenna element to another, the circuitry will simply choose again.

Phased Arrays A much more sophisticated way of using multiple antenna elements is to construct what is called a *phased array*. Here the outputs or inputs of the various elements are constructively or destructively merged to form beams of almost any desired shape, and this can be done at both the transmitter and receiver. Beams can even be steered over many degrees of arc to direct energy off to one side.

Phased arrays have existed for decades, but until rather recently they were manually configured, and the phase relationships were relatively fixed. Much more recently, adaptive array antennas, or *smart antennas*, have been developed that utilize a computing engine to shape beams dynamically on a channel-by-channel basis in order to concentrate energy in whatever direction and at whatever intensity is desired. Some such antennas can even steer the beam dynamically and track a moving object, thus providing the receptor terminal with a strong direct signal at all times.

Adaptive Array "Smart Antennas" Essentially two types of adaptive array antennas have been developed: the switched-beam antenna and the beam-steering type. The switched-beam antennas can combine only the beams from the different element in a finite number of juxtapositions. The beam can assume a few fixed widths and a few fixed angles and is not infinitely variable. The beam-steering type is, on the other hand, infinitely variable and is far more flexible. It also requires far more processing power to operate effectively.

Smart antennas excel in NLOS applications for a number of reasons, not all of which are pertinent to all designs. Most significantly—and this does apply to all design variants—the antenna array has the ability to focus a beam very tightly toward each subscriber unit on a packet-by-packet basis. The RF energy in that beam is not dispersed through the atmosphere as in a normal broadbeam transmission and instead is delivered almost in its entirety to the subscriber site, where it can blast through considerable obstructions and still provide a usable signal. By concentrating energy in such tight patterns on a channel-by-channel or even packet-by-packet basis, the adaptive array can also increase frequency reuse, theoretically up to several times within a single cell, while still adhering to regulatory limits on transmitter power. This in turn reduces the need for sectorization.

All adaptive array antennas provide signal diversity; that is, the mere fact that several spaced antenna elements are exposed simultaneously to the transmission practically guarantees that signal quality will vary from element to element. The system can then select the best signal, and in an NLOS situation, there is a far higher likelihood that an array will find a usable signal than a single element. In some systems the adaptive-antenna array can even take multipath reflections impinging on the various elements and phase-align them so as to construct a single, coherent, high-strength signal.

Other systems can combine various signals and then use vector-cancellation strategies to reduce interference. In one experimental system developed by Lucent but never marketed, the smart antenna system could simultaneously receive several signals over the same channel, separate them on the basis of time of arrival and multipath signature, and then reconstruct usable signals out of all of the interfering transmissions, thus reusing spectrum at virtually the same point in space! Figure 5-6 shows an ArrayComm adaptive array antenna.

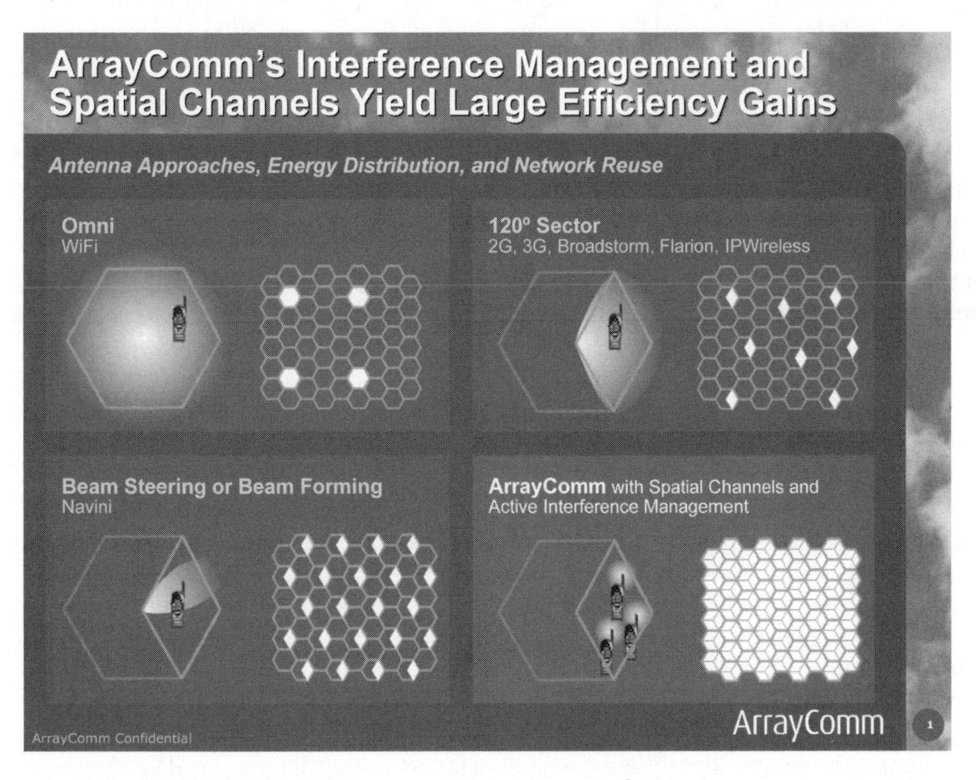

Figure 5-6. *ArrayComm adaptive array antenna, courtesy of ArrayComm*

Adaptive array antennas are most beneficial when used in both the base station and the subscriber terminal. Such double-ended systems are known as multiple-in, multiple-out (MIMO) links (the multiples refer to the antenna arrays). Currently there is no explicit MIMO standard for 802.16, though a standard is in preparation for 802.11.

I recall first reporting on such intelligent phased arrays back in 1996 and hearing such capabilities discussed even then, and, predicting—erroneously, as it turned out—that intelligent adaptive array antennas would sweep the wireless industry. I also attended a number of demonstrations of advanced adaptive array technology and can attest that it brings real benefits that should be appealing to network operators.

Why, then, has it not in fact become established in the marketplace?

In a word, price. Adaptive array antennas require multiple radios, one for each antenna element, and heavy-duty processors. To date, most have been based on expensive field programmable gate array circuitry rather than application-specific integrated circuits (ASICs), which further drives up the price. Although the capabilities of the network are most enhanced when adaptive array antennas are used at both ends of the airlink, the high price of such antennas has ruled out their use in subscriber terminals thus far, and thus the full potential of the technology has seldom been realized. At least one company in the broadband wireless business, IP Wireless, promises to introduce moderately priced adaptive array antennas for subscriber units some time in 2004, but, if it occurs, this will be a first.

At the time of this writing only a handful of companies manufacture equipment utilizing adaptive array antennas intended for fixed broadband wireless deployments. These include the aforementioned IP Wireless, Navini, BeamReach, Redline, Orthogon, Vivato, and ArrayComm. Of these, only Orthogon makes a MIMO system, and that system is intended solely for point-to-point wireless bridges. Vivato's equipment is really designed only for 802.11 networks, and ArrayComm makes a component antenna module, not a complete system. Thus, current choices for the network operator seeking a networkwide single-vendor deployment of adaptive array antennas really only number four. Interestingly, of that four, two, Navini and IP Wireless, make products supporting mobility.

Despite the small number of vendors today, adaptive array antenna technology is almost inevitable since it permits much more efficient use of spectrum than any passive antenna system. By the end of this decade it will be the norm, and equipment manufacturers without the technology will not be competitive.

At the same time, I am not suggesting that any of the existing systems featuring adaptive array antenna systems are necessarily the best choice for any given network operator in any given market. Adaptive array technology comes at a cost premium, and it is for the network operator to decide whether the increased access to potential customers will result in enough additional revenues to balance the higher equipment costs. One should also keep in mind that NLOS capabilities are only one attribute to consider in choosing infrastructure equipment. One also has to consider the routing and/or switching options built into the base station, additional support for quality of service (QoS) beyond the 802.16 standard, built-in support for voice, and so on.

My advice is that any broadband operator should look at adaptive array antennas closely and attempt to test equipment. In other words, it is a choice that should definitely be explored. Whether it is to be embraced at this time is another matter.

Mesh Networks and NLOS

Chapter 3 has already discussed the principles of mesh network operation fairly extensively. This section focuses on the NLOS capabilities of meshes and how a mesh architecture relates to the previously mentioned techniques.

Using a mesh topology does not exclude any of the methods for facilitating NLOS mentioned earlier. A mesh network can use OFDM and/or adaptive array antennas, and in fact one system manufactured by Radiant Networks (now out of business) did utilize a rather crude mechanically steered antenna that was the functional equivalent of a smart array in certain respects. Indeed, if one were to combine a smart antenna with a mesh, one would achieve a powerful synergy, but so far no manufacturer has attempted this.

By itself a mesh offers a compelling solution to NLOS problems in that it allows the network to route around obstacles—provided enough subscriber terminals are present to permit circuitous routing paths. The drawback, of course, is that the multiple hops required to describe such paths rob the network of capacity and can introduce latencies that compromise time-sensitive applications.

As indicated earlier, the future of the mesh in public networks remains in doubt. The companies advocating the architecture are, with one exception, all very small and may not be able to survive in the overall broadband wireless marketplace where there have been too few deployments to date to support a large and diversified population of equipment manufacturers. And there are also issues involving the installation of the terminals that could compromise the effectiveness of what is essentially a sound networking strategy.

In a mesh, as you have seen, the subscriber terminal and the base station are one and the same. This means that the radio and antenna are likely to be self-installed and the antenna placed indoors at low elevation. The result will be that the individual radio will be provided with a number of possible transmission paths but none that are likely to be particularly good, especially, if, as is the case with most current mesh systems, the antennas are omnidirectional. Such antennas also give rise to a generally high level of background interference because the signal is being propagated everywhere. But without adaptive array antennas, or physically steered antenna such as Radiant makes, an omni is the only way to ensure that the signal can reach multiple nodes in the network.

The choice of a mesh over a point-to-multipoint architecture is fundamental and will affect every aspect of network performance and evolution afterward. It is a choice not to be undertaken lightly. My view is that mesh products are less proven than point-to-multipoint but may be indicated in certain circumstances. If, for instance, the network is being planned for a dense urban core where tall buildings ring every conceivable base station location and prevent the establishment of links with choice customer locations, then one may want to explore what may be done with a mesh, in particular whether line-of-sight connections can be established among initial nodes that would enable future prime customers to be easily reached. Unfortunately, this exercise will have to be done manually, at least in part, because not many software tools are designed for planning meshes. And, in any case, the fact that mesh equipment itself is designed to be self-organizing rather than operator configurable limits the degree to which the network can be planned.

As is the case with any other radio, the operator would want to examine the mesh radio in terms of its other capabilities—can it support QoS, voice telephony, legacy protocols such as Asynchronous Transfer Mode (ATM), and so on? NLOS is of little use if one cannot offer the services and reliability the customer expects.

Adaptive Modulation and Cell Planning

Adaptive modulation is a term denoting the ability of a radio to change its modulation scheme on the fly to adapt to varying signal conditions. In existing commercial products the radio is not going to shift from CDMA to OFDM or anything that fundamental. Instead, the choice will probably be between quadrature phase shift keying (QPSK) and 64 quadrature amplitude modulation (QUAM) or perhaps 16 QUAM.

Here I will provide only the briefest definition of the two basic techniques.

Phase shift keying is somewhat akin to frequency modulation. The phase of the carrier wave is retarded or advanced by so many degrees, and the phase shifts are used to represent bits. By providing, for example, four different degrees of phase shift, the signal can carry 4 bits of information per hertz or wave cycle, though with normal losses this rate is unlikely to be consistently realized in practice.

QUAM adds amplitude gradations to those of phase—in other words, the intensity of the carrier wave as well as its phase position is made to vary by so many discrete steps, and the more steps, the more bits can be encoded in a single waveform. In theory hundreds of bits per hertz could be encoded by this means, but with normal losses and overhead involved in radio transmissions, 5 bits per hertz is about the maximum throughput that can be achieved today.

The greater the number of states that can be represented by a single waveform, the smaller the differences between states and the greater likelihood that the encoded information will be lost to interference or noise. Thus, QUAM modulation systems can only be used when signal strength is very high.

A radio capable of adaptive modulation will constantly evaluate signal quality and will shift from high-order QUAM to lower-order PSK when a more robust signal is needed. The throughput rate will drop accordingly. In some cases adaptive modulation can be overridden in a radio, but the network operator can generally expect higher bit error rates when that is done.

The effects of adaptive modulation on throughput rates, especially over distance and in the presence of multipath environments, must be taken into account when the network is being planned. Generally, adaptive modulation will degrade capacity while improving signal integrity, but the operator needs to know exactly to what degree for each within each individual cell and sector.

Frequency-Agile Radios and Network Mapping

Chapter 3 also alluded to frequency-agile radios. I am reasonably confident that they will play a major role in the networks of the future. They will certainly increase the capacity and spectral efficiency of any broadband wireless network in which they are used, but they will also add enormously to the complexity of cell mapping and network organization. I know of no network planning software capable of modeling networks based on this facility, so the operator today has no real means of planning for the arrival of a technology that is three to five years off in a commercial sense. My assumption at this point is that frequency agility would only be an asset when combined with adaptive array antennas and with the overall network intelligence that would enable every base station to coordinate its activities with every other base station and solve extremely complex problems concerning frequency allocation and reuse within the entire network.

The Installation Process

I have already discussed the issues of determining where to place base station and subscriber nodes. Here I will briefly examine the actual process of mounting the equipment.

Mounting the antenna is the most critical part of the installation process. As you have seen, antennas are best installed at considerable elevations, minimally 25 or 30 feet above street level and preferably much higher, and the longer the distance the transmission must reach, the higher the elevation generally. Antennas may be installed on utility poles, roofs, specially constructed towers, and even hilltops. In all cases, the antenna should be kept well away from reflective surfaces.

In many instances, a rooftop installation will require a steel supporting structure, and this must be securely fastened to a stable surface so that the antenna assembly can withstand wind loading and earthquakes. This same structure must also be properly grounded to endure lightning strikes.

Most cities of any size boast antenna specialists with deep experience in installing base station antennas, and such individuals should be utilized by the wireless operator either on a contract basis or as regular employees. Haphazard installation of terminals will compromise the reliability of the network considerably and may expose the network operator to heavy liabilities if equipment comes loose or causes a fire.

If the radio itself is installed outside, it must have a case specifically designed to withstand any conceivable adverse weather condition, one that will not permit condensation to form within the case. National Electrical Manufacturers Association (NEMA) certification provides a good indication of a piece of equipment's ability to withstand the elements, and the network operator should insist on it for equipment that is to be installed outdoors.

If the antenna and/or radio is to occupy a utility pole, the installation crew should work closely with the utility in question and make certain that the installation is done in compliance with local ordinances and the utility's own internal procedures. Rules are not uniform in this regard, so the network operator should not act upon assumptions.

Incidentally, poles carrying power lines are not good candidates for installation, both because of the physical danger they themselves represent to the installer and because they radiate very strong magnetic fields that can interfere with sensitive radio circuits.

Proper cabling procedures are equally as important as the physical positioning and securing of the base station antenna and radio. Poor connections can impose unacceptable losses on the signal and lead to complete disruption of the service. Fiber-optic connections, where they are present, are even more critical because cut fiber strands must be very precisely aligned to pass a signal at all. When fiber connections are required, they should be made by trained installers with proper cable splicing equipment.

Frequency Converters, RF Amplifiers, Integrated Radio/Antennas, and Radio Modems

The network operator faces certain choices as to how to configure equipment used in a base station or subscriber terminal.

The first choice in regard to a base station is whether the radio will be combined with the antenna. Such combinations make for generally simpler installations and allow the installer to run a data stream on ordinary data cable back to a router or switch. The drawback is that a limited amount of functionality can be built into such a compact unit. If, on the other hand, a feed must be taken from the antenna to a transceiver, then the signal quality is more likely to be impaired. In some cases where a radio/antenna combination is used, the installer may choose to employ a downconverter that will shift the transmission to a lower frequency where it is less subject to attenuation within the antenna cable.

The second choice is whether to use an outboard amplifier. Radio frequency amplifiers are devices for boosting a signal that are sometimes added to transceivers in base stations. In the United States they are forbidden for use in the unwired spectrum unless they have been specifically designed around a given transceiver. Outboard amplifiers add to the cost of the system but may be necessary in some instances where the transceiver simply lacks the power to reach certain subscribers.

Subscriber premises equipment is generally simpler than that at the base station, but, in some instances, the radio may also be part of the antenna. In other cases the antenna lead goes to an outboard modem, often called a *brick* in the trade. The tendency today, especially in indoor installations, is to place the modem and the antenna on a PCMCIA card that can be installed directly in a computer's card slot.

Signal Distribution Within the Subscriber Premises

In the case of a single residential user, the transceiver can communicate directly with a single computer via whatever connection the radio and the computer itself supports, be it USB, serial, or FireWire (IEEE 1394). In the case of enterprise users or multidwelling unit (MDUs), generally some provisions must be made for distributing the signal after it has been received over the airlink. What this entails is the creation of an internal network for the building or compound in question. Figure 5-7 shows an installation in an MDU.

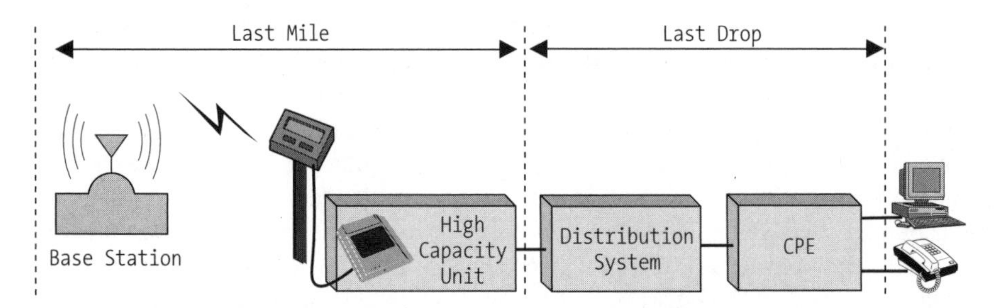

Figure 5-7. *MDU installation, courtesy of Alvarion*

Here network operators will have to decide to what degree they want to become a building local exchange (BLEC) and enter into the business of installing and maintaining LANs. In some cases having this capability will mean the difference between securing an account and failing to do so, but remember that LAN installation at its higher levels involves competencies that are somewhat remote from those needed to set up and operate public networks, and this is especially true if extensive new in-building wiring is required.

Internal Networks for the Enterprise User

A few words of explanation are in order regarding enterprise in-building and campus data networks; however, taken has a whole, enterprise networks represent a vast subject area and cannot be adequately covered in a single section. Here I am primarily interested in how such LANs interface with wide area networks (WANs) and with metro area networks (MANs), the latter being the real province of the broadband wireless service provider.

First, an enterprise network proper—a network that would be used within an institutional setting such as a corporation, government agency, or an educational or research institute—is a rather different entity than a distribution system intended for an MDU. The latter ordinarily has no switching capabilities (that is, the ability to define Ethernet subnets), though it may have rudimentary routing abilities primarily involving dynamic Internet Protocol (IP) addressing with perhaps some virtual private network (VPN) functionality thrown in. An enterprise network, on the other hand, can be extremely complex and can include its own application and content servers, special black boxes for conferencing, hardware firewalls, authentication servers, and on and on. Normally such networks will be managed by an information technology (IT) department, and the WAN or MAN service provider will play no direct role in respect to what goes on within the LAN itself.

Things get more complicated still when several tenants within a single facility share the same internal physical layer, that is, when several enterprise LANs share the same wires or internal airlinks. At that point, the internal network looks more like a public network because some means of ensuring fairness must be implemented, and data streams and databases associated with individual customers must be encrypted and made inaccessible to other tenants. In other words, the network must be actively managed. Usually this will be the responsibility of the building owner, but in the past a number of specialized service providers, the aforementioned BLECs, have attempted to specialize in setting up and operating these internal networks. The MAN service provider can elect to function as a BLEC, and a number have done so.

Dedicated BLECs have generally failed in the data communications marketplace, often because of exorbitant demands on the part of real estate owners for shares of the revenue. Consequently, the wireless network operator who is contemplating filling such a role should give the matter careful thought. If a large number of valued customers reside in one location, it may be justified, but only under certain circumstances.

Generally, fulfilling a BLEC function is advisable only in the case of millimeter wave services where a connection of minimally several hundred megabits per second serves a single building or complex. In such a case enough bandwidth exists to provide several customers with several tens of megabits of throughput. The internal network can be either IP or switched Ethernet and normally would utilize either Gigabit Ethernet copper data cable or optical fiber. Chapter 6 gives detailed information on the setup and administration of both types of networks.

MAN/LAN Integration: The Physical Layer

In any given instance, the network operator will face the choice of whether to offer a separate wireless subscriber terminal to each subscriber in a building or attempt to serve such tenants through an internal network spanning the last few yards. Wired connections may appear to be the more sensible approach for the final connection—after all, cable is pretty inexpensive—but the issue is not always as straightforward as it may appear.

First, as you have seen, the installation or modification of internal cabling thrusts the wireless service provider into a new role that of the BLEC. At the point when the signal enters a wired network and network operators are responsible for its delivery over that network, they have to utilize new tools and techniques that do not obtain in the outdoor wireless environment. In effect they have entered into a different business.

Here cost is often the overriding consideration. Cable installation is fairly inexpensive in buildings under construction, especially where the installer enjoys a good working relationship with the general contractor. Adding new wiring to existing structures can be both expensive and laborious, however, and simply may not be tolerated by the building owner.

Determining a Business Strategy Regarding the Internal Network Most larger office buildings are wired for data today, and some of the biggest even have internal optical fiber. Many have internal routers or Ethernet switches as well for distributing data services to tenants. The more elaborate the internal network, the more likely the building is to have a fiber connection.

In the case of large office buildings and complexes, the network operator's role will depend on many factors: the extent of the physical plant already existing in the building(s), the posture of the real estate owner, the network operator's own expertise, and the way that the operator has positioned the company in the marketplace. Each installation is unique, and there are really no invariant step-by-step procedures to guide the service provider through the process.

If the structure to be served has no cabling or inadequate cabling, then it is probably best to enlist the cooperation of an engineering firm specializing in such work, of which there are many. Large office buildings can require months of installation labor to cable, and total costs may run into the hundreds of thousands of dollars. Incidentally, traditional telecommunications service providers have tended not to involve themselves in designing, building, or operating internal networks.

It is of course possible to distribute signals internally via 802.11 WLANs, and constructing such a network should be well within the capabilities of a wireless public network operator. The problem is capacity. Specifically, 802.11 networks using direct sequence modulation can accommodate only three users per base station simultaneously, and throughput speeds are 108Mbps best case. There are certain proprietary "Turbo" accelerators that might enable speeds close to that but they are not part of the standard.

In the future ultrawideband radios may solve this internal distribution problem in a cost-effective manner. With throughputs in the hundreds of megabits per second, and pronounced ability to penetrate walls, they could eliminate the need for much of the internal cabling that festoons modern office buildings. Currently, however, the power restrictions in force in the United States will not permit such applications.

Another option, alluded to briefly earlier, is to provide each subscriber in a building with a separate radio and antenna. This is best attempted in the lower microwave region because of the relatively low cost of terminals and their NLOS capabilities.

Even if new cabling is not required, new networking equipment may be, such as edge routers, Ethernet switches, digital private branch exchanges (PBXs), firewalls, and so on. And these in turn may require new equipment closets and, in some cases, whole data centers. Thus, it is extremely important to determine the real market potential of each building and have firm agreements in place with real estate owners before attempting to make any modification in an internal network. It is equally important to develop good estimates of the cost involved in such modifications. A network operator simply cannot afford to guess wrong in this regard. To do so could jeopardize the whole network operation.

Interfacing with Enterprise Networks

Enterprise LANs vary in size and complexity from small assemblages of a few networked computers to sprawling networks with hundreds or even thousands of terminals. The larger networks may include IP PBX functionality, hardware firewalls and authentication servers, multimedia servers, internal routers and switches, IP or Ethernet network attached storage, and so on. Dealing with such formidable entities, especially as a startup service provider, may seem daunting, but the very complexity of the network may actually represent an opportunity for a competitive service provider.

In reality, many telco incumbents do a poor job of serving enterprise accounts. Most traffic over incumbent public data networks still takes place over legacy transports, including SONET, frame relay, and ATM, and it involves complex and inefficient interfaces to the enterprise world, which is overwhelmingly Ethernet and IP. Enterprise data managers routinely complain that incumbent business services personnel do not really understand their networks and are not responsive to their needs. And, up to the present, the telcos have not had to be responsive since they enjoyed a de facto monopoly.

Generally, providing the enterprise IT manager with a fat IP pipe with self-provisionable bandwidth will go a long way toward satisfying that individual. Ensuring QoS over long-distance remote connections will do even more.

To achieve the latter, the wireless service provider has to enter into special relationships with long-distance carriers that will ensure that certain stated levels of speed, latency, and error rates will be maintained such that service level agreements with the broadband customer can be maintained. Especially important here is minimizing the number of router hops to the destination and assignment of special priorities to certain kinds of traffic. All this costs money because the wireless network operator is asking for more than basic Internet service, but that cost can be passed on to the business subscriber, and the cost will generally be justified from the latter's perspective if the service enables new applications that benefit the enterprise.

Such long-distance peering relationships, as they are called, are especially important if the service offerings include videoconferencing or IP voice services. They are equally important if highly specialized services such as links for real-time collaborative scientific computing or massive multimedia file transfers for entertainment industry production houses are offered.

If the wireless broadband operator is offering network storage services, special provisions will probably be necessary as well, specifically the installation at the central office of specialized equipment to handle the service, either dedicated storage switches or multiservice switching platforms that can handle storage transport protocols.

The network operator may also be obliged to invest in dedicated encryption hardware when dealing with enterprise customers. The security protocols inherent in 802.16-based equipment may not be deemed sufficient, and the customer may demand something more. When bulk encryption of highly sensitive transmissions is required, hardware solutions are preferred because they utilize a platform that is entirely separate from the router to perform the computations necessary for encryption and thus do not slow it down. Chapter 7 discusses encryption in detail.

Another specialized service offering that should be considered with respect to enterprise customers is access to a secure Web site that will permit the subscriber to change the throughput rate or other attributes relating to QoS. Most enterprise users have occasional needs for large increments of bandwidth, and naturally they do not want to sign up for increased throughput rates for periods longer than are absolutely required. Access to a self-provisioning Web site answers such needs.

A word about legacy networks: A significant though dwindling number of enterprise networks retain networking protocols that are currently considered legacy. These include ATM to the desktop, token ring, Novell NetWare, AppleTalk, Banyan, and X.25, to name the best known. While many first-generation WLAN products supported such protocols, few current broadband wireless products do, and thus integration with such networks may represent considerable difficulties. Accordingly, the network operator must carefully weigh the value of potential customers demanding support for such protocols. Are they worth the operator's retaining staff or consultants with expertise in these areas?

Internal Networks and the Residential and SOHO Subscriber

Interfacing an 802.16 subscriber premises radio with a residential data network ordinarily poses few problems. The overwhelming majority of such networks are Ethernet or, in some cases, small IP networks fronted by routers.

The situation is rather different when the subscriber wants a network, but the house has not been prewired to support it. In that case the broadband wireless network operator may want to consider offering basic installation services as a value-added service.

In such cases, putting in an 801.11 WLAN is usually the simplest option while installing category-five data cable is the most difficult, expensive, and time consuming. Other options that do not require running new cable are HomePNA, a sort of mini-DSL network that utilizes the house's existing telephone wiring, and HomePlug, a standard for running high-speed data transmissions over the AC power lines.

In the case of MDUs where one radio may serve a number of subscribers, the installation and management of the network can become much more difficult. The 802.11 standard, HomePNA, and HomePlug may still suffice if the network were limited to a handful of subscribers, but if the number is greater than that, the network operator may want to establish subnets where contention for bandwidth is limited to the subnet. Home PNA and Home Plug are not really designed to support Ethernet switching and the establishment of subnets, but interestingly, 802.11 is.

As I have indicated before, MDUs do not appear to be an ideal market for wireless broadband operators to pursue, based on return-on-investment projections. Millimeter microwave connections are still prohibitively expensive in view of likely revenues, and lower microwave networks generally cannot provide much capacity and are therefore at a competitive disadvantage vis-à-vis cable and DSL. Moreover, cable and DSL service providers are relieved of the necessity of constructing separate internal network to redistribute the signal. DSL requires no redistribution, and in the case of coaxial cable the external network naturally lends itself to extension within a building. Generally an MDU customer should be actively pursued only where other broadband services are lacking and are unlikely to be offered in the near term.

Infrastructure for a Purpose

If one had to formulate a single rule for the deployment of physical infrastructure in a wireless broadband network, or any other type of broadband network for that matter, it would be the following: Every aspect of the deployment must harmonize with an explicit service model. Cell location and dimensions, frequency reuse patterns, and modulation techniques must all support subscriber demographics and the various service offerings at predicted usage levels. Constructing a network is as much an exercise in market projection as it is a technical undertaking. When that is understood by the operator, the considerable risks associated with any broadband access business are significantly reduced.

CHAPTER 6

Beyond Access

The previous chapter focused heavily on the physical infrastructure of the broadband wireless network and how it is put in place. This chapter concerns itself with the higher layers of the network having to do with the management of network traffic and the differentiation of services. In other words, the focus is more on what the broadband wireless network has in common with other high-speed access networks than on what sets it apart.

The Place of the Central Office in the Business Case

Chapter 4 briefly discussed the physical requirements for a central office facility in terms of the space itself. This chapter considers how the central office figures into the profitable administration of the network, precisely what network elements are likely to occupy the space, and how these elements relate to higher-layer networking functions.

The Role of the Central Office

Above all, the central office is a network hub—in point-to-multipoint networks it is a larger hub aggregating the traffic of the smaller hubs that are the individual base stations. It is also the point where the broadband wireless network connects to the public switched telephone system (PSTN), the public Internet, and, in some cases, satellite networks. It can also connect to remote storage networks, though as yet wireless broadband has played a small role in the storage services industry.

The central office is much more than a relay point for inbound and outbound network traffic, however. It is also generally a repository for customer records, including billing information; the place where the entire operation is administered; the point where subscriber authentication occurs; the nexus for service creation (that is, defining value-added services for specific subscribers); the place where specialized content is cached and distributed; and the home of any network security appliances in use in the operation.

Central Office Traffic Management and Distribution

As indicated earlier, wireless networks can utilize a number of transport protocols for handling traffic to and from the subscriber premises and wide area networks (WANs), but the 802.16 standard is based on using Internet Protocol (IP). Given that, routing is the dominant traffic distribution mode within the network.

IP in its basic form is a pure packet protocol, which requires using a router or packet switch. This is a device that transmits packets (that is, discrete and discontinuous sequences of data) as capacity becomes available in the network. During periods of congestion, packets are held in buffers until an opportunity to transmit presents itself. Packets themselves are of variable length, and the number of bits in each packet will vary according to network conditions.

By packetizing a signal rather than transmitting it in a continuous stream over an open circuit, the router tends to use bandwidth much more efficiently than a circuit switch. A switch must hold a circuit open even when a delay in the data stream occurs, while a packet router will allow other traffic to occupy the open bandwidth. Such mingling of various messages within the same channel tends to degrade transmissions that require the maintenance of precise timing such as voice and particularly video, however, and this is the penalty that packet networks customarily exact upon the user.

Each packet is provided with an address to enable the router receiving the transmission to sort out the various messages. This address is quite distinct from the uniform resource locator (URL), or Web address, and is normally unseen by the subscriber. In essence, it is the province of the router and is utilized to plot routing paths. Since the supply of Internet addresses provided by the dominant IP 4 is not inexhaustible, the usual practice is for a broadband network to obtain a few permanent IP addresses, which interface with the outside world, and rely on internal addresses to communicate with subscribers from the base station. The permanent IP addresses are also known as *global addresses* because the entire universe of Internet routers can see them, and the internal addresses are called *local addresses*. The outside party communicating with the subscriber ordinarily sees only the permanent address and not the address assigned to the individual user. Network address translation (NAT) from a global to a local address takes place within a router maintained in the central office.

In some cases, particularly important customers such as large enterprises or government agencies will be assigned permanent IP addresses, and indeed many will insist on this prerogative. The wireless network operator planning to solicit such customers should be aware of their requirements in this regard and should obtain a sufficient number of addresses to accommodate them.

Another method of dealing with a shortage of Internet addresses is Dynamic Host Control Protocol (DHCP), which distributes addresses on a dynamic basis among subscribers. The whole process takes place transparently and automatically. A global Internet address is "leased" to a user for a predetermined length of time, which could be as short as the transmission itself or weeks or months. DHCP is really a form of oversubscription, or statistical multiplexing, enabling the network operator to get by with less than a single address per user on the theory that all users will never be online simultaneously.

DHCP is usually administered from a separate server, not from the router itself, and the addresses themselves will normally be assigned to enterprise users who would probably own internal routers of their own.

Incidentally, IP 6, which has not been widely adopted yet in the United States, has an address space sufficiently wide enough that no shortage of addresses is anticipated even if separate devices within the network such as computer peripherals and smart appliances are assigned their own addresses. Whether the computing community at large will ever embrace IP 6 remains to be seen.

Routers

Routers themselves come in several forms that are determined by the function of the device in the network. *Core routers* are huge, costly devices placed at major Internet access points and hubs and handle the enormous volumes of long-distance Internet traffic. *Edge routers* are much smaller devices used in metropolitan access networks that constitute "the edge" from the perspective of the long-distance service providers. Yet smaller routers are located within enterprises, and the smallest of all are located in residences and small businesses. Most wireless broadband base station controllers made today happen to incorporate edge routers. To see how routers fit into a public network, refer to Figure 6-1.

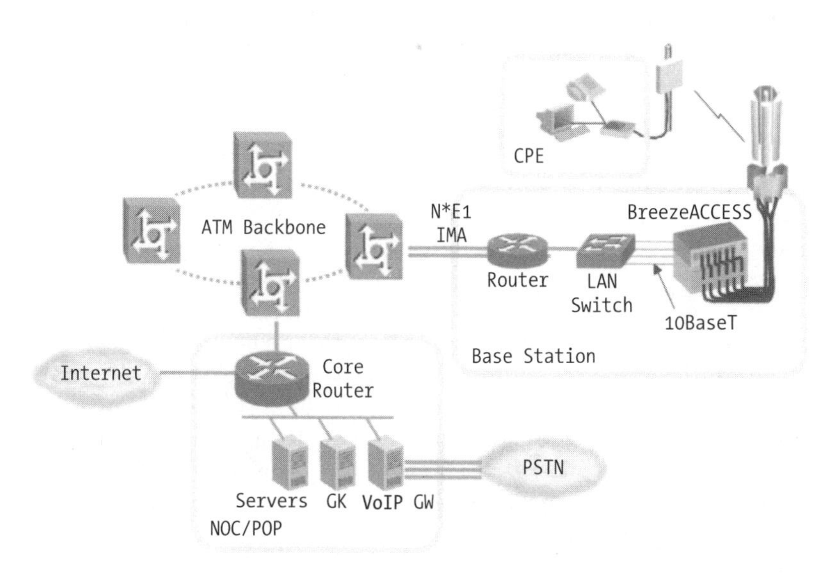

Figure 6-1. *Wireless network with router, courtesy of Alvarion*

Routers at base stations are switches, but they frequently perform much more complex functions than traditional class 5 and class 4 circuit telephone switches. Whereas circuit switches simply follow instructions and send messages to the next node in a predetermined network path, routers may be capable of making instantaneous decisions as to an overall route through the Internet based on communications from other routers. Thus, one packet may take a different route from another, and the packets may arrive at their final destination out of sequence, whereupon they will be held in a queue until all the packets have arrived. Information in the packet header will enable the destination router to assemble them in correct order.

Routers make traffic decisions based on routing tables that are continuously and automatically updated in large core routers and in the edge routers utilized in metro networks. These tables list locations of core routers across the Internet and the regions served by them, and they function somewhat analogously to the numbering system used in the PSTN.

Dynamic routing of this sort customarily takes place in edge routers and core routers but frequently not in enterprise and residential routers where static routing schemes tend to prevail. The presumption is that the configuration of the local network will be a given and that network traffic patterns will be entirely under the control of the local area network (LAN) administrator.

Routers were initially designed to switch text data and still-frame graphics only (in other words, information that could tolerate the delays inherent in packet transmissions). Today routers are conveying all sorts of delay-sensitive traffic including but not limited to voice; video, including high-definition video; high-fidelity multichannel audio; interactive entertainments; video surveillance; video conferences; and business voice-over applications such "white boarding" and real-time, online, agent-assisted sales presentations.

For a router to provide good presentation quality with such delay-sensitive content, two things generally have to happen. First, the content to be transmitted has to be cached as close as possible to its final destination. Second, the router itself has to begin to operate more like a switch.

The latter generally involves a new packet-switching protocol that I have mentioned before known as Multiprotocol Label Switching (MPLS). Whole books have been written about MPLS and the various aspects of the protocol, and I will not attempt to explain it in detail here. My aim is more modest, simply to indicate what MPLS means to the broadband wireless operator.

MPLS is based on a concept initially known as *tag switching* that has been around for almost a decade and was first associated with Cisco Systems and Toshiba, both of which developed prestandards router/switches with tag-switching capabilities. The idea was that the basic IP router functionality would remain intact—no separate and divergent packet-switching protocol such as frame relay was envisioned. Instead, a switching function would be built on top of an IP such that at least certain classes of traffic would take determinate paths through the network—paths that could involve reserved bandwidth. These paths would be designated by the tag or label associated with the traffic in question, and that traffic, instead of being routed through first one path and then another according to network conditions, would take but a single path. Each router/switch would strip the label from the designated stream of data and rewrite a new label, indicating its next destination. There would be no consulting of routing tables, no lookup, and no path determination procedure. In the case of the label-handling function, the router would instead function essentially as a dumb switch.

According to MPLS, labels are assigned to traffic on the basis of common parameters such as permissible latency, committed bit rate (if any), maximum jitter or timing errors, and so on. Flexibility in provisioning bandwidth to meet service requirements far exceeds that associated with asynchronous transfer mode (ATM) or frame relay, the legacy packet standards for handling diverse types of network traffic.

MPLS itself is not nearly as comprehensive as ATM, the protocol it most resembles, however. MPLS is basically a technique for segregating and shaping traffic according to its need for controlled bandwidth, but unlike ATM, which performs much the same function, MPLS lacks internal mechanisms for reserving bandwidth or enforcing network performance to ensure that service levels are maintained. These are provided by other protocols such as the Resource Reservation Protocol (RSVP), used for bandwidth reservation, and DiffServ, which, as the name implies, posits instructions for differentiating various services. MPLS merely segregates the differentiated traffic into streams and attaches the appropriate labels for handling that traffic in transit.

MPLS is embedded in most edge routers sold today and in nearly all new core routers. It is not common in the combined base station controllers/routers used in broadband wireless networks, though. True, all equipment that is 802.16 compliant will have quality of service

(QoS) mechanisms that are inherent in the standard, but these should not be considered a complete substitute for MPLS. The routing provisions specified in 802.16 are nowhere near as comprehensive and flexible as those in MPLS, and, moreover, they are not intended to operate end to end over WANs. At best, they are a substitute for older ancillary standards for QoS such as DiffServ and RSVP built around wireline networks.

The question still remains as to how quickly and how extensively MPLS will penetrate the metropolitan area network (MAN). Like ATM before it, MPLS ultimately rests upon the assumption that rich multimedia offerings will play an ever-increasing role in broadband access systems and that IP will become the preferred transport protocol for delivering such services. Given the slow but steady growth of animated graphics and streaming audio and video over the Internet, it does seem reasonable to suppose that multimedia will assume much greater prominence in the future. But what is absent is any clear indication that the telecommunications industry is approaching a turning point. Internet television is still largely confined to short news clips, movie previews, and music videos, and, despite the efforts of companies such as Intertainer to introduce comprehensive video services to broadband operators of IP networks, such offerings remain curiosities. Internet radio has done better, if for no other reason than it is easier to achieve acceptable presentation quality with audio-only programming over medium-speed connections—which is what most so-called broadband services are today. Even so, Internet radio has not proven to be a major moneymaker for the service providers that carry it.

I fail to see a clear path ahead for IP multimedia services; the growth of such services is almost inevitable, but the lack of distinctive or successful program offerings to date is troubling. Much of the problem obviously has to do with bandwidth and resolution. Even with the most up-to-date compression technologies, minimally about a half megabit per second throughput is required for good video quality. When the IP video transmission has to compete with channelized high-definition television programming over cable, satellite, and terrestrial broadcast, a 120-kilobit stream full of drop-outs and motion artifacts probably is not going to attract much of an audience. Recently, a number of companies such as Microsoft, Myrio, and Minerva have demonstrated high-definition video transmissions over IP networks, and such demonstrations are encouraging. In the near term, however, only the DSL service providers are likely to offer high-definition IP television.

Nevertheless, I see MPLS-switching capabilities as an investment in the future. By acquiring an edge router that can handle MPLS traffic, network operators will be ready for the new services when they emerge. They will also be better able to ensure QoS end to end in LAN extension applications across WANs.

Ethernet in the Metro and the Advisability of Placing Ethernet Switches in the Central Office

Most of the wireless broadband equipment adhering to a metro Ethernet model conforms to the 802.11 standard rather than to 802.16. The 802.11 equipment is Ethernet based, and 802.16 is IP based; that's the distinction.

Ethernet is a layer-2 protocol rather than layer 3; in other words, it employs circuit switching to direct traffic over the network instead of packet routing despite that Ethernet itself is a pure packet protocol. Switched Ethernets divide the overall network into subnets, generally based on the geographical location of the terminals, and the switch sends traffic to the appropriate subnet. To see how Ethernet switches fit into a public network, refer to Figure 6-2.

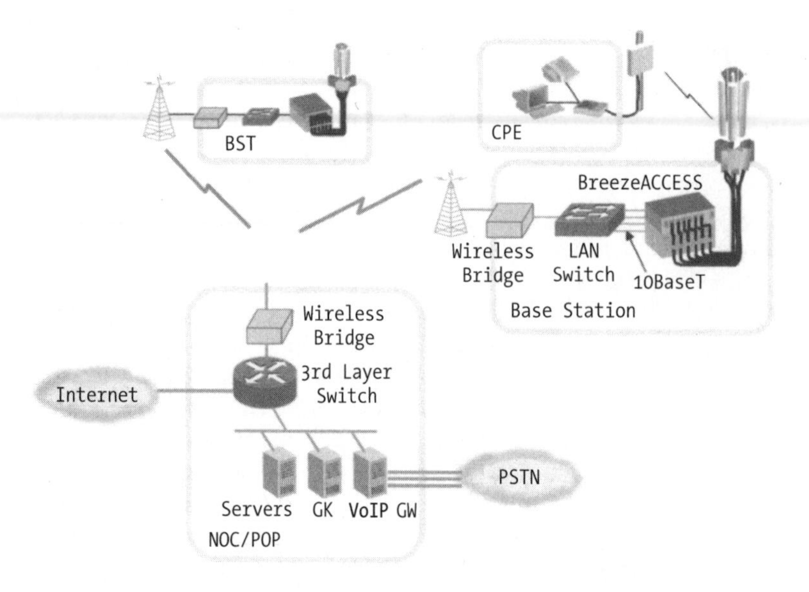

Figure 6-2. *Wireless network with Ethernet switch, courtesy of Alvarion*

I am mentioning Ethernet switches here in the context of the central office because a considerable body of opinion says that Ethernet rather than IP is the proper protocol for the MAN. In fact, no fewer than two industry associations, the Metro Ethernet Forum and the Gigabit Ethernet Alliance, have arisen to promote such views.

Within the telecommunications world, a high degree of partisan fervor often accompanies advocacy of one protocol or the other, and indeed the stakes are high and are nothing less than market success or failure for the party embracing one standard or other. A dispassionate consideration of all relevant transport protocols is far preferable to rabid partisanship, and the criteria on which a transport should be judged are the cost effectiveness and flexibility of the equipment in supporting a given service model as well as the likelihood that the standard itself will gain wider currency in the future.

That said, you should consider the past and future of metro Ethernet: High-speed metro Ethernet business class services were introduced by several competitive data services in 2000, including offerings from Yipes, Cogent, Terabeam, and CAVU-eXpedient, among others, the last two of which used wireless infrastructure. The initial acceptance for such services was encouraging, but two, CAVU-eXpedient and Yipes, declared bankruptcy, and neither Cogent nor Terabeam has succeeded in posing major challenges to the frame relay and T1 incumbents, though they have managed to survive. Recently large long-distance companies as well as local incumbent telephone carriers have also begun to introduce both IP and metro Ethernet services, though as yet such services have very small subscriber bases.

The independent metro Ethernet service providers all followed a similar model, offering simple high-speed access at high throughput rates and no value-added services whatsoever. Basic Ethernet is well suited to such a model, but it is much less suitable to a more contemporary model where the network is used to deliver converged services. An Ethernet substandard for supporting virtual private networks (VPNs), 802.1q, does exist, as well as another, 802.3x, which imposes flow control over the network. But far less standards work has been done for the purpose of enhancing network performance with delay-sensitive traffic than in the case of IP.

The general Ethernet solution for any problem is to throw more bandwidth at it rather than to invoke sophisticated network management mechanisms, which explains the steady and rapid increase in speed from 10 megabits per second (Mbps), to 100Mbps, to 1 gigabits per second (Gbps), and finally to 10Gbps. Incidentally, a 100Gbps standard is in preparation.

Some companies, most notably Extreme Networks, have developed extensive proprietary QoS protocols for Ethernet that greatly augment the capabilities of the Ethernet switch, but one is still left with the question, why not use a standard-based approach such as IP/MPLS?

In a book of this sort, any exhaustive review of the arguments pro and con for Ethernet in the metro would be inappropriate. The fundamental argument is that Ethernet-based components will be cheaper because of mass production of products intended for the enterprise and residential markets. Certainly, some of the same chipsets can be used in carrier products as in less-demanding applications, but a degree of redundancy and intelligence must be provided in a product intended for a public network, and such requirements will serve to elevate the price. The relatively few metro Ethernet products to appear thus far have not been inexpensive.

A second argument has it that enterprise managers, the individuals who will actually be using broadband services, are happiest when interfacing to an Ethernet that is simply a larger version of what they themselves manage. There is some truth to this argument. But given the lesser capabilities of Ethernet in the area of service creation vis-à-vis IP/MPLS or ATM, is this singular virtue sufficiently compelling?

My view is that the momentum behind metro Ethernet has significantly lessened. Furthermore, only two companies, Atrica and Extreme Networks, currently manufacture pure carrier-grade metro Ethernet switches (numerous companies make multiservice-switching platforms that can do Ethernet switching as well as routing). The whole notion behind the original metro Ethernet movement of simply offering high-throughput speed to a subscriber and leaving the subscribers to establish their own services strikes me as largely discredited today. Still, one never knows. Perhaps some service provider will make a success with this model yet.

Also, an Ethernet switch can be combined with an MPLS switch with or without including full IP routing capability. Products from Atrica and Extreme Networks are essentially MPLS-enabled Ethernet switches, and Laurel Networks, TiMetra (now a division of Alcatel), and Vivace (now a division of Tellabs) make boxes that can function as Ethernet switches, IP routers, and MPLS switches. None of these products is specifically designed to operate within a wireless broadband network, though, and interface problems may arise. Incidentally, all these companies are startups that tout at least partially proprietary approaches to service delivery, and all their products are expensive and intended for inclusion in larger networks.

Finally, the inclusion of an Ethernet-switching capability at the central office may render a wireless service offering more attractive to certain customers. Here again, it is good to assess the needs and profiles of potential subscribers before proceeding with the construction of the network.

VPNs and the Central Office

VPNs are enabled internally by most carrier-grade edge routers and Ethernet switches today and by many base station controller/routers offered by broadband wireless manufacturers. Using such a centralized platform rather than component encryption devices makes for easier administration, a smaller footprint in the central office, and generally lower equipment expenditure. I strongly believe in MPLS- or Ethernet-enabled VPNs over older implementations involving tunneling protocols and extraneous boxes.

Telephone Switches

Telephone services were commonly offered by the first-generation millimeter wave wireless operators but seldom to date by networks operating in the lower microwave regions. Chapter 7 covers voice telephony and the protocols and equipment required to implement it, but here I will venture a few words as to how telephone switches figure into the central office environment.

In traditional local exchanges, the switch was the primary piece of hardware occupying the central office. In broadband networks where the emphasis is on data, switches are of secondary importance and are lacking in many networks.

As I stated earlier, I do not think provisioning circuit telephone services is a good business for a wireless broadband operator to enter, particularly a startup with limited financial resources. Class 5 circuit switches cost millions of dollars, and, considering that most wireless broadband networks start small and remain fairly small, acquiring at most a few thousand subscribers, paying millions of dollars for a device that is likely to yield a few dollars per subscriber per month simply does not make economic sense—not in a highly competitive service environment where infrastructure equipment has a far shorter life span than in the past. The full return on investment for a class 5 switch would require decades, by which time it would be completely obsolete. Also, given that telephone switches are not designed to interface with radios and that complex procedures for mapping voice channels onto airlinks, providing dial tone and ring tones, and assigning telephone numbers are required, the prospect of doing circuit voice appears distinctly uninviting—except perhaps in developing nations in regions where wireline telephone services are completely unavailable (such fixed wireless telephone services are known as *wireless local loop*). Cable operators, for reasons best known to themselves, have frequently offered circuit voice, and they have had to put in complete cable telephony hardware and software platforms to enable such services as well as the aforementioned class 5 switches. If anything, circuit over broadband wireless is even more problematic.

Softswitches

IP telephony is a different matter. Equipment costs thousands rather than millions of dollars, and many manufacturers are already building broadband wireless radios with inherent support for IP voice. Therefore, the barrier to entering the IP voice market is much lower than is the case for traditional circuit voice. Still, it is a business not to be embraced without careful consideration and one that demands hard choices in respect to basic infrastructure equipment.

Two fundamental equipment-related issues confront the broadband wireless operator considering IP voice services. The first has to do with the choice of a platform, and the second has to do with the scope and positioning of the services.

IP telephony today suffers not from a lack of standards but a plethora of them. H.323, Universal Datagram Protocol (UDP), Session Initiation Protocol (SIP), Simple Object Access Protocol (SOAP), Media Gateway Control Protocol (MGCP), and Media Gateway Control (MEGACO), among others, vie for the allegiance of the IP voice carrier, and in no case are these all supported on a single hardware platform. IP telephony also suffers from a diversity of approaches in embodying those several standards in hardware. Often, though not always, signaling and converting IP traffic into circuit traffic takes place within different devices, and some manufacturers favor highly distributed architectures where numerous aggregation boxes are scattered about the network. In the case of SIP-enabled networks, special telephone instruments are used that convert the voice signal to IP right at the deskset. Each approach requires a different type of deployment with a different array of equipment components.

Wireless broadband operators face a choice of terminating local phone traffic over their own networks at either a class 5 switch or a softswitch equivalent or, alternatively, ceding the switching function to some other service provider. In most cases, the ownership of a traditional class 5 switch is out of the question for the wireless operator, so a softswitch is the only real choice unless the wireless operator chooses to route phone traffic through the exchange of the local incumbent and utilize that carrier's circuit switch.

If a softswitch is employed, the wireless operator will need to lease a circuit connection to a class 4 or tandem switch operated by a long-distance service provider. Both long-distance calls and local calls made to parties who subscribe to landline telephone services will then have to be routed through that class 4 switch. The local calls to nonsubscribers will then be sent back to the incumbent telco's class 5 switch and from there to the designated party. Obviously, some complex issues and difficult decisions are involved in setting up an IP voice service.

Quite a number of service models can be built around IP telephony, and I will discuss them in detail in the next chapter. The long-term profitability of those models is at issue, however. Voice is becoming an increasingly price-eroded, low-margin business, and IP telephony will surely accelerate that trend because it reduces the cost of transit and because IP voice has been positioned as a price-competitive product by those carriers offering it. To make matters worse, various peer-to-peer Internet telephony programs have recently been disseminating through the online community that completely eliminate the need for a local carrier and that route all telephone traffic through the Internet service provider (ISP) with no billable minutes being involved. IP voice services may be becoming cheaper to provision, but they also appear likely to command lower prices over time and ultimately lower margins for the service provider. Some industry analysts have concluded that voice telephony will eventually become a no-charge amenity that is simply included in a basic service package. Figure 6-3 shows a voice-over IP connection to the PSTN.

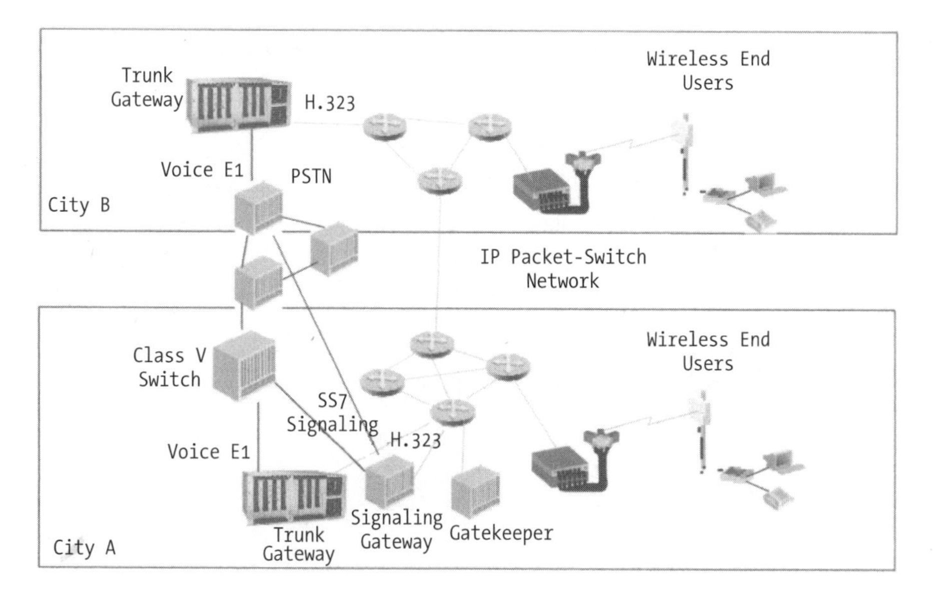

Figure 6-3. *Voice-over IP, courtesy of Alvarion*

If such a state of affairs comes to pass, then voice services could become just another overhead expense for the wireless broadband operator. But, at least for the midterm, offering voice still appears to represent a sound strategy for the wireless operator, if for no other reason than it adds to the bundle of converged services.

Before I leave the topic of telephone switches in the central office, I will add this concerning the issue of availability: Conventional plain old telephone service (POTS) has come to be regarded as a lifeline service, something that will enable subscribers to summon aid in the midst of emergencies, and indeed countless lives have been saved over the years because of timely telephone calls. Naturally, the value of a telephone service as a lifeline depends on availability, and at least in the United States, incumbent carriers have made every effort to ensure the highest availability. One attribute of traditional analog phones that makes such availability possible is that no outboard electrical power source is required. As long as the telephone is connected to an active jack, it will operate even during a power outage.

In general, this is not true with broadband voice services such as voice over digital subscriber line (DSL), cable telephony, or broadband wireless voice. They utilize AC or battery power to operate the interface and thus are vulnerable to outages.

The cable industry has responded by designing customer premises telephony interfaces with backup power, but I have not seen anything equivalent in the wireless broadband space. For that reason operators should position their voice offerings as second-line services. I also suggest that any IP switches be highly redundant and served with instantaneous backup electrical power.

Mobile Voice Equipment

Mobile voice can be supported on some 802.16-compliant equipment, but at present no one makes an 802.16 phone. IP Wireless, a broadband wireless equipment manufacturer that makes a cellular telephone–like high-speed infrastructure platform, currently supports voice as well as high-speed data, however. Any mobile voice offering over broadband wireless could not be done through a simple overlay because of the requirement for base station handoffs but would require a base station controller, such as the IP Wireless product, designed from the ground up to perform handoffs.

ATM and Frame Relay–Switching Equipment

The 802.16 standard supports ATM as well as IP, and at least one broadband wireless manufacturer, Alvarion, supports frame relay. To that extent, ATM- or frame relay–based services are an option for the broadband wireless operator. Normally, to provision either service, a separate dedicated switch is required; however, a number of so-called godbox multiservice-switching platforms are on the market that handle both types of traffic within the confines of a single box. For an illustration of how an ATM switch fits into a wireless public network, refer to Figure 6-4. This architecture characterized the first generation of broadband wireless services and is still permitted in the 802.16 standard.

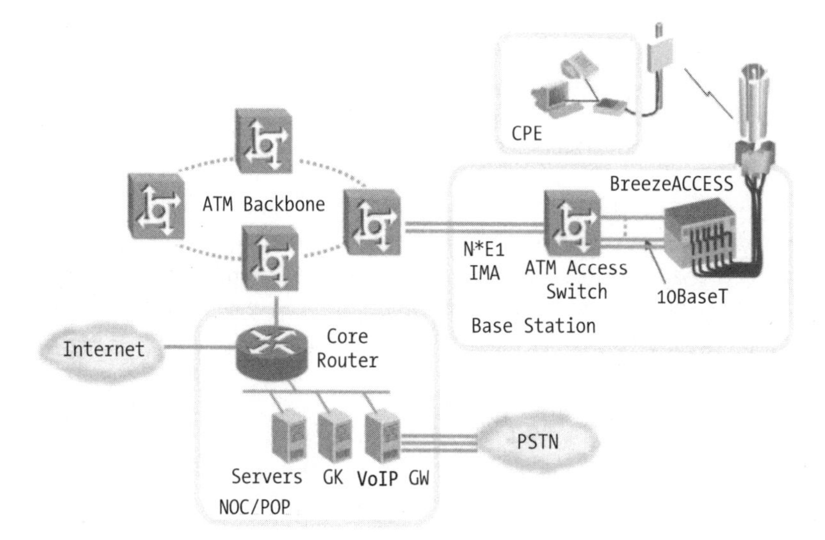

Figure 6-4. *Wireless public network, courtesy of Alvarion*

Investing in an ATM switch or frame relay access device (FRAD) is generally not warranted in a broadband wireless network today. Frame relay services are legacy and relatively inflexible compared to IP since support for multimedia is not well developed. ATM, while it provides excellent QoS for a wide range of services, is bandwidth inefficient. Moreover, the ATM at the desktop enterprise equipment that would justify the use of this protocol at the central office has ceased to be manufactured. Within the 1998–1999 time frame the possibility existed that ATM would indeed become the universal platform that its creators always intended it to be, but the extremely high cost of the equipment precluded significant penetration in the enterprise, let alone in the consumer residential market. ATM continues to be widely used in long-haul networks and will be for many years to come. It is also extensively used in DSL aggregation and, somewhat curiously, in the FireWire connections used in consumer multimedia applications. Furthermore, most first-generation Local Multipoint Distribution Service (LMDS) millimeter microwave equipment was ATM based, as was some first-generation lower microwave point-to-multipoint hardware. But beyond these established niches, ATM is not experiencing growth or expansion. In any case, ATM switches remain extremely expensive devices priced in the tens of thousands of dollars. I cannot conceive of how such an allocation of resources could possibly be judicious in a metropolitan broadband network.

Application-Specific Equipment for the Central Office

Directing network traffic is the primary function of the central office, but it is far from the only function. The central office also houses equipment that supports specialized services and applications.

What this equipment will be depends on the suite of services the operator has elected to offer. I have discussed some of the value-added services in Chapter 3, including mobile voice, conferencing, telemetry, storage, and entertainment multimedia content. One can envision further applications, including interactive video commerce, such as has been done to a limited extent in some European broadband networks; Web hosting, which is normally the function of large specialized ISPs; Web mirroring, a subset of the latter; and so-called walled-garden content that generally consists of local advertising and news items, which are not available over the public Internet.

Conferencing equipment requirements vary according to whether the conferencing takes place over IP, ATM, or Integrated Services Digital Network (ISDN). ATM-based video conferencing has been the gold standard for the industry, but it is expensive to support, both for the subscriber, who requires a special hardware component, and for the service provider. ISDN is an obsolescent data standard that was commonly used for videoconferencing services in the past despite the fact that it is expensive and difficult to administer. The trend today is to migrate toward IP.

IP conferencing may involve special dedicated hardware components or simply software installed in the subscriber's computer. When it is offered as a fully supported service by a broadband access provider, it usually involves not one but several dedicated hardware components.

A complete discussion of conferencing will follow in Chapter 7. Here I will deal only with the central office requirements for the service.

In an IP videoconferencing implementation within the central office, the service provider will normally purchase equipment that can also be used for ordinary IP voice telephony; in other words, it will be dual use. A gateway—generally occupying a separate computing platform—will mediate between the metro network and the WAN and in most current equipment will allow a video conference to take place across protocols. In other words, the individuals in the broadband wireless network using IP-based equipment can conference with someone in a remote location having only an ISDN connection. A separate component called a multiconferencing unit (MCU), which is a kind of specialized switch that allows any number of callers to participate in the conference, is necessary when more than two parties are involved. Finally, a gatekeeper handles the administrative functions associated with the conferencing session, including billing. The gatekeeper software may have its own dedicated computing platform or may simply be a blade in a server running other operations support systems (OSS) software.

To maintain high image and sound quality, the IP transmissions must take place over a network that is MPLS enabled from end to end. Thus, the broadband operator will have to put in place agreements with long-distance service providers that will ensure that such arrangements prevail.

Videoconferencing is a rare offering among wireless broadband operators, but it is within the capabilities of an 802.16 network. In my view it is an attractive value add that will be increasingly in demand in the future.

Storage Networking Devices

Storage services are provided through what are known as *storage area networks*. I do not know of a wireless broadband service provider that has offered such services, but the services are certainly possible over a wireless network. As indicated in Chapter 3, storage services often involve special transmission protocols, such as Fibre Channel and ESCON (although IP and Ethernet storage networking products are achieving increasing acceptance). I know of no wireless broadband equipment that provides for native support for any storage networking protocol.

In the past, service providers that offered storage services tended to define the offerings in one of two ways. They would offer a complete outsourced storage service where data backup in remote data centers would be provided, as well as the download and retrieval of stored data over a high-speed network, or they would simply provide a transport for an enterprise to transfer data to a storage site of its own choosing. Companies offering complete storage solutions have not fared well in the marketplace, and today storage transport appears to be the better service offering.

Storage transport normally requires a separate storage switch, though some multiservice platforms can handle storage protocols as well as IP and/or Ethernet. Enterprise-class storage requires quite a bit of bandwidth and thus would be best attempted over millimeter microwave or aggregated unlicensed spectrum in the 5GHz region. I suggest that any operator contemplating offering storage services first poll customers and prospective customers to determine the likely take rate of such a service before buying equipment to support storage applications. To date, storage networking has not been a big business from the carrier perspective, though that could change.

Web Hosting and the Central Office

Web hosting is a service generally provided by specialized ISPs, not by local broadband access providers. I know of no instance of a broadband wireless network providing this service, though it is certainly feasible. Web hosting requires what is known as a *server farm*, which is just what the name implies—a data center holding numbers of servers supporting numbers of Web sites and Web-based broadcasts, multicasts, unicasts, and transactions. Often, though not always, the Web hosting company will form a relationship with a specialized provider of transport such as NaviSite, which will ensure the expeditious delivery of hosted content.

Web hosting is a complicated business that is really separate from the high-speed Internet business. As with broadcast video, it demands a fundamentally different kind of physical plant (although the two could be combined, albeit at considerable expense). Web hosting is best left to specialists.

The MMDS broadband wireless services began as networks dedicated to delivery of video, and some local networks still function in this manner. A few have even been successful, primarily in remote areas where a well-established cable incumbent is absent, although today with the ready availability of direct satellite broadcast services almost everywhere it is doubtful how long the traditional wireless cable model will remain viable.

Also, Multichannel Multipoint Distribution Services (MMDS) convergent services networks have been set up on the cable television model where a couple of 6MHz channels have been allocated to residential high-speed access while the rest of the spectrum has been assigned to television channels (though I know of no outstanding successes based on this approach). Given the relatively limited amount of bandwidth available in the MMDS bands, the model seems somewhat dubious, though it could work with digital television equipment where several channels apiece are assigned to the 6MHz frequency slots. The problem is that most such networks have been built on the macrocell model and cannot reuse the data channels, and therefore no considerable population of data users could be served.

My belief is that the wireless cable model is wrong for most markets today. In any case, the limited bandwidth with which to work would appear to confront the network operator with a crucial choice: to specialize in one-way video programming or to specialize in two-way high-speed data. The two approaches require different equipment complements within the central office. The equipment requirements for data services are discussed at length throughout this chapter and do not differ significantly in terms of higher-layer networking equipment from what one would find in a DSL network. A video-oriented system, on the other hand, requires what is essentially television station hardware, including video tuners, up and down converters, satellite feeds, ad insertion servers, scramblers, video archival hardware, and so on. Given the poor record of wireless cable plants in making profits, I do not see why anyone in a developed market would try to resurrect this approach.

I have alluded earlier in Chapter 3 to the possibility of doing video on demand (VOD) over a wireless broadband network. No one to my knowledge has done so, but techniques have been developed for VDSL networks that could easily be applied in the wireless realm and would not require the reallocation of vast amounts of bandwidth. Such an approach would require the purchase of a dedicated video server, a device that may range in price from thousands to tens of thousands of dollars depending on port count, and, if continuous programming were provided, banks of tuners that subscribers could access to select individual programs. I would hasten to add that at this time VOD is a completely unproven service offering for wireless broadband, but the possibility exists that it may find a market in the future.

Somewhat related to VOD in the sense that VOD is essentially one of its subsets is *interactive television*. Again, this represents a market that is embryonic in the wireless realm but that could conceivably emerge in the future and certainly bears watching.

Interactive television consists of programming that invites the direct participation of the viewer. Examples include the following:

- Participating in t-commerce, where a viewer can click an item in an infomercial or even an episodic program and then call up a Web site with product information and the capacity for credit card transactions
- Participating in game shows such as typing answers to a question or voting
- Selecting viewing angles during a sporting broadcast
- Ordering takeout food off an on-screen menu
- Summoning video catalogs containing desired products
- Calling up news clips of events of interest

Interactive television applications have enjoyed some success in European cable television networks but have been generally rejected within the American market to date. Currently no specialized equipment exists for their delivery over wireless broadband; within cable networks, specialized set-top boxes and headend servers are required to enable such services. There does not appear to be any pent-up demand for such services, but the market could change.

Successful network operators have to be able to anticipate changes in markets, and for this reason close monitoring of interactive television is indicated at present. Clearly, any early entry into this market with improvised equipment is not advisable, though.

Specialized Content Distribution Platforms

This section refers to a rather amorphous category of service offerings having to do with either content that is associated with an event such as online show or exhibition that may be streamed to a subscriber or content that is local and is made available only through the individual broadband access network. The latter is known as a *walled-garden* service.

Video streams often involve satellite downloads to avoid a trip through the public Internet and the subsequent caching of the content onto video servers where said content is then streamed over a high-speed last-mile connection. The technology for doing this is fairly well developed and could be adapted to a high-speed wireless networks, but the services to date have not been big moneymakers over wireline networks where they have been tried, and broadband wireless operators should adopt a wait-and-see attitude.

Walled gardens have not proven particularly successful in broadband access networks either, though the cable networks particularly have long anticipated that such services would ultimately prove profitable. Set-top box and cable headend manufacturers have built walled-garden capabilities into both customer premises equipment and cable headend systems. The content itself will normally be stored on a special server. To date no one has made equipment having a walled-garden capability that is aimed at the wireless broadband market, and such services represent a potential rather than an actual market.

Location-Based Services and the Central Office

Location-based services can scarcely be said to exist today within the wireless broadband space. Essentially these require terminals that will announce their geographical position to the base station, which will then consult a database with a profile of each user. Content such as local advertising can then be pushed at the user or made available on demand. Ordinarily a separate server would be required to support such services.

Location-based services have aroused the greatest interest in mobile networks where localized information is often derived from Global Positioning System (GPS) coordinates, and the mobile terminal itself contains a miniature positioning system. To date, these services are virtually nonexistent in the United States, but they have been tested in Europe.

When broadband wireless networks acquire a full mobility component, they may begin to offer location-based services, the assumption being that both technologies may mature at the same rate. For the time being, such services remain speculative.

Hotspot Services: Central Office Requirements

Normally 802.16 would not be used to deliver hotspot services directly to a customer; rather, the connection to the customer would occur over the 802.11 wireless LAN (WLAN) protocol, and an 802.11 base station aggregating hotspot traffic may then utilize 802.16 for backhaul to a central office.

Hotspots represent a fairly new service category, and little uniformity exists in terms of the equipment used to enable them. In the case where a broadband wireless metro service is deploying the hotspots, the usual practice would be to place the customer database, authentication function, and billing within the central office. The authentication would most likely take place within a separate authentication server, but the billing and administrative software could reside on the same servers used by the network at large. Separate billing software would probably be required because of the prevalence of roaming and credit card transactions in hotspots.

OSS and Network Management

OSS software is a topic treated at length in Chapter 8. OSS and other management software would normally occupy one or more servers within the central office, with authentication usually being assigned to a dedicated server. Customer databases may be assigned to yet another dedicated server. To ensure network reliability, both the OSS and the databases should be mirrored, particularly anything having to do with billing. Backup storage will, of course, be a necessity. In most cases, one would probably opt for what is known as *network-attached storage* rather than remote storage since the central office itself would be designed as a secure facility. Storage would usually take the form of redundant arrays of magnetic discs and would involve special controller elements for distributing data redundantly through the array. Considerable floor and rack space may be consumed by such arrays, and sufficient allowances should be made for storage when designing the central office.

Security Devices and Appliances

Security involves primarily two types of devices that can be deployed in the central office: firewalls and encryption boxes. Chapter 9 describes the functions of both at length.

Firewalls regulate traffic going into a network and are meant to keep out intruders and unauthorized users who are attempting to assume the identities of legitimate users. The intent is to prevent parties from outside the network from seizing control of computers attached to the network and examining their contents, using them as platforms from which to conduct illicit acts, or attempting to sabotage the computers. *Stateful inspection* firewalls examine IP addresses and other aspects of incoming traffic and determine whether the sender has any business being in the network. *Proxy server* firewalls move transactions to servers mediating between the subscriber's computer(s) and the WAN so that the subscriber's databases and software cannot be directly accessed. Usually, a stateful inspection firewall running on its own physical platform would be the device used to secure a public network. Proxies tend to slow down network traffic and would require the network operator to mirror a considerable number of subscriber computers in the network.

Authentication servers, mentioned in the preceding section, are sometimes considered to be security devices and sometimes OSS platforms. They determine whether parties attempting to use the network are really who they claim to be and thus regulate access to the network.

Yet other security platforms run network diagnostics and counteract attacks on the network. These tend to be used more in the enterprise than in public networks.

A large part of security has to do not with hardware but with the proper setup and administration of a network. One wants to be careful as to allowing remote access to the operating and administering systems of key network elements such as routers, switches, gateways, and so on, and also to servers hosting OSS or customer databases. It is also a good idea to encrypt any really vital data pertaining to the customer or to network statistics and financial records. Public networks have been hacked in the past, and they will be in the future. A successful hack that succeeds in shutting down the network or requires the reconfiguration of routers and servers could put network operators out of business and could expose them to heavy liability. For this reason, proper security measures should be an integral part of normal network operations.

Beyond the Central Office

Within the telecommunications business, no common carrier or diversified service provider can exist in splendid isolation. An isolated network is like the old narrow-gauge railways built in mountainous areas that could not connect to major railways; a communications networks unconnected with the outside is just as limited.

The following sections explore the various relationships that a broadband wireless service provider must forge with other service providers in order to serve the needs of customers. Certainly some service providers can be nothing but competitors, but many relationships between network operators are mutually beneficial.

Ultimately, setting up a successful broadband wireless access involves a great deal more than erecting base station towers and signing up subscribers, although certainly both processes are essential. It involves more than establishing link budgets for all the airlinks and frequency coordination throughout the network, though these tasks also are essential. Preeminently, the network operator has to decide how desired content and applications are going to be moved across the network, and that will be determined by the nature of the connections with other networks.

In many cases, a metro network with a wireless broadband component will also have wireline elements. The wireless network may extend the reach of a cable television, a DSL network, or even a fiber-optic network, or the network may be truly hybrid, with wireline and wireless parts coexisting in a rough parity. In either case, the wireless portion of the network must be joined with one or more wireline segments.

Many 802.16 base station controller/routers have DSL and fiber-optic ports to facilitate such connections, but if the network operator is actually operating rather than interconnecting with a wireline network, a complete complement of appropriate hardware devices will be required. In the case of optical networks, optical transponders and optoelectonic converters will be needed, and in the case of DSL, DSL modem cards and aggregation devices will be necessary; however, obviously, any discussion of the intricacies of DSL, hybrid fiber coax, or optical network operation is far beyond the scope of this book.

Essentially, all such networks would be unified with a wireless broadband network through the higher layers of the network, primarily at layer 3, the layer where IP routing takes place, since all these systems would be accessible via routers. In most cases, direct translations from one physical platform to another where routers are bypassed are not possible.

In some cases, the wireless broadband network operator will have no choice but to use a synchronous optical network (SONET) interface to off-load traffic onto a SONET ring for transport to an Internet exchange point or a class 4 or 5 telephone switch (or both). Such equipment would usually be installed and owned by the service provider selling the connection to the ring. IP and Ethernet connections to optical rings are also possible and are cheaper and simpler to install and administer, but they are still fairly uncommon in most optical rings, because, for good or ill, SONET still remains the established standard.

Broadband Wireless Networks in the Larger Context: Connecting to Other Service Providers

For those of whom the following information is obvious or superfluous, I apologize, but no one should finish this book without a firm grasp of what follows, whether acquired here or elsewhere.

The Universe of Networks

The world today is served by a vast number of different electronic communications networks accessible to the public. Some of these exist in a hierarchical relationship to one another while others are autonomous or semiautonomous. What is important to recognize is that the latter—namely, the more or less isolated networks—are coming to be viewed as increasingly anachronistic. Where in the past networks such as cable television systems and two-way dispatch radio networks were truly autonomous, now both are likely to be connected to the Internet and to the public switched telephone system, and thus ultimately to one another. Even paging networks tend to have Internet and telephone interconnectivity today.

Currently, two all-embracing networks exist, the PSTN and the Internet, which is the universal network of data networks. The Internet, back in the days when it is was known as ARPANET and was designed primarily to serve the needs of the military and other government agencies, was conceived as an overlay on the PSTN, which would utilize many of the Bell Operating System network elements already in place. For this reason the two cannot readily be distinguished or disentangled from one another.

The traditional PSTN is in the process of gradually morphing into a sort of "super Internet" or next-generation Internet, a process that has crucial implications for everyone in the broadband access today—or in any other aspect of telecommunications, for that matter. When that process is completed, the procedures of connecting one physical network with another, such as a wireless broadband network with a fiber ring, will become simple and straightforward. Moreover, the role of every business entity owning physical infrastructure is apt to change as a result, in some ways that are still unpredictable but in others that are apparent even now (I will discuss these ways in later sections). For now, functional and structural distinctions continue to exist among networks, and it is to these that I will now turn your attention.

The PSTN: The First Hierarchy

The PSTN gradually assumed its present form over a period of more than 100 years, but almost from the first it began to exhibit its core architecture. This was based on a point-to-multipoint topology on the local level, with lines extending out from a local exchange, sometimes called the *central office*, and proceeding to larger regional aggregation points that themselves were linked together in a mesh arrangement that spanned the continent and ultimately the world. For decades all connections, both short and long, took place over copper wire and involved mechanical switches that moved pieces of copper wire. Furthermore, all communications took place in the analog domain.

During the 1960s, fully electronic switching came into being while voice signals began to be digitized. Otherwise, however, the physical plant remained much the same.

During the last two decades, the physical plant of the PSTN has undergone considerable modification. The tendency today is to terminate copper-pair phone lines at what are called *digital loop carriers (DLCs)*, which are neighborhood aggregation points, often mounted outdoors and utilizing hardened enclosures. The analog voice signals carried by the copper lines will then be digitized and will be conveyed back to a central office over optical fiber. Dial-up Internet connections will undergo a similar transformation inasmuch as they are actually carried within analog audio signals.

DSL lines will terminate at a DSL access multiplexer (DSLAM) and will generally also run over fiber from the aggregation point to the central office. A DLC for ordinary phone lines and a DSLAM can and often do occupy the same enclosure today.

Today the central office itself is a facility that is occupied by a device known as a *class 5 switch* whose purpose is to manage traffic over circuit voice channels. The switch's primary function is to establish a path to the call's intended destination utilizing available network resources, but it must also be capable of handling a multitude of special features such as call waiting, call forwarding, conferencing, and so on.

Other special calling features, such as caller ID and one-number portability, are handled by special applications servers that are dedicated to those functions coming under the rubric of Advanced Intelligent Network (AIN). Basic switching functions are compromised when the switch's control plane is invoked to perform unrelated tasks; hence, the use of separate platforms to support special features has nothing directly to do with traffic handling.

Telephone switches, incidentally, are enormously complex devices embodying millions of lines of code. Interestingly, most of the functionality enabled by the code occurs beneath the surface. Such complexity is reflected in the price, which runs into the millions of dollars.

The class 5 switch is involved in directing calls within the local exchange itself and outside the local exchange to other local exchanges, which could be located anywhere in the world. The immediate path that the signal takes is most likely to be onto what is known as a *fiber ring*, however. The class 5 switch itself does not handle call routing to other class 5 switches. That function is performed by a class 4 or tandem switch, described next.

Fiber rings emerged in the 1980s and followed the SONET standard in the United States and followed the closely related synchronous digital hierarchy (SDH) standard in most other places in the world. The rings themselves always take the form of closed loops, but they are not necessarily rings in the strictest sense and may in fact meander over the landscape to a considerable degree. The reason a closed-loop architecture is used is to provide an alternative path back around the loop if a break occurs in the fiber.

Although some fiber rings encompass only a single metropolitan area and primarily distribute data services to businesses within an urban core, the larger rings extend out over a greater metropolitan area or even several such areas; the area may enclose thousands of square miles. These link many local exchanges and ultimately off-load their traffic onto larger switches known as *class 4 switches* that handle calls going to other area codes or country codes. In some cases, long-distance carriers own the larger rings, but in other cases independents such as American Fiber own them. The rise of independents in this area constituted a major trend in telecommunications during the late 1980s and early 1990s. Some cable operators also own fiber rings—generally the smaller, metro variety.

Fiber rings themselves are evolving, and one tendency that is beginning to manifest itself—though more in the East Asian markets than in the United States—is the adoption of packet protocols intended to replace SONET. Chief among these is Resilient Packet Ring (RPR), which combines the fast restoration capabilities of SONET with the spectral efficiency of IP and Ethernet. Such packet rings are capable of handling all types of traffic, including voice.

Class 4 switches located within the rings send traffic out over *long lines*, which in the past were copper or microwave links but today are almost exclusively optical fiber. The long lines in use today consist of hundreds of strands of optical fiber, each capable of carrying more than 100 frequencies of light. Each frequency in turn can convey minimally 10Gbps or as much as 40Gbps. Thus, total capacity is in the terabits. Such lines are generally the province of long distance or inter-LATA (which stands for Local Access Transport Area) carriers, as they're known in the business, which form the second or third rank in the greater telecommunications hierarchy depending on whether one designates metro or regional rings as a separate stratum.

At the top of the hierarchy are companies such as Tyco Telecommunications and Global Crossing, which command international and transoceanic fiber. These companies came to prominence during the late 1980s and through the 1990s. The submarine cables they utilize are typically of very large capacity, equivalent to that of the largest continental long lines, and are designed to provide years of maintenance-free operation.

This neat and somewhat idealized hierarchy has been considerably confused by the emergence of the Internet, however, as I will explain in the next section. But before that, I will cover the role that this first hierarchy, the PSTN, will play in the business of the broadband wireless service provider.

If one wants to provide voice telephony services, even IP voice services, then one must in most cases either link up with a central office class 5 telephone switch via a device known as an *IP voice gateway* or purchase a class 5 switch of one's own. This is because ultimately the call is going to a telephone number somewhere else, not to an IP address. This means that the broadband wireless operator will have to form a relationship either with the incumbent telephone carrier or with a facilities-based competitive local exchange (CLEC) such as a cable operator that happens to own a class 5 switch.

In fact, a number of long-distance services use IP as a transport, but, ironically, most of them also utilize gateways to translate the voice traffic back into circuit form. It is possible to transmit a call end to end to an ordinary telephone deskset entirely over IP networks, but it is not the norm today. One would have to establish a special relationship with a long-distance IP voice service to do so.

Incumbent telcos have offered access to the larger PSTN at a price to other local service providers for a long time; two-way radio dispatch services are a prime example. If the number of voice customers in one's network is small, simply leasing a single T1 back to the incumbent's central office may suffice.

In the case of voice telephony services, one always must be concerned with being beholden to one's competitor, and any arrangement with an incumbent local exchange (ILEC) or even a CLEC tends to put one in that position. One possible solution, at least in the case of facilities-based CLECs, is to attempt to make the relationship synergistic. Some cable operators, for instance, are starting to view wireless broadband as a means of reaching certain customers for data services who are not passed by cable currently and cannot be cost-effectively served by this means. Independent mobile operators are another possibility since they invariably own class 5 switches. A relationship in which the broadband wireless operator provides access to such customers under some revenue-sharing arrangement in return for a connection to a class 5 switch at a reasonable rate could be advantageous to both parties.

The Public Internet: The Second Hierarchy

The Internet, as you have seen, is built largely on basic telecommunications infrastructure. All, or almost all, signals pass over the same basic physical infrastructure of last-mile twisted-pair copper and metro and long-haul fiber as do voice calls, but once onto long-haul fiber or, in some cases, metro fiber, ordinary circuit telephone switches do not determine the direction of the data traffic any longer. Rather, the switch sends the data traffic to a *network access point (NAP)* in what is usually a local telephone call, and from there circuit-switching procedures of the sort utilized in traditional telephone networks are no longer invoked. Instead, a *core router* assumes the function of directing data traffic through the network.

Core routers are specialized switches that switch individual packets rather than complete streams, and, in the case of best-effort traffic without stringent QoS requirements, the individual packets may take different routes through the network to avoid congestion.

Routers themselves may be yoked with ATM switches or may even incorporate ATM functionality, and in such cases the ATM protocol may be used to encapsulate IP traffic. This results in some loss of efficiency but, for reasons that go beyond the scope of this book, aids in shaping traffic and managing bandwidth. I see ATM gradually losing ground to MPLS, which has similar capabilities but offers greater efficiency and much more flexibility when it comes to service creation. The tendency today in the core routers used at major Internet hubs is to combine the functionality of an MPLS switch with an Internet router in a single box.

NAPs and their associated routers are generally owned and operated by long-distance carriers, though a number of independents exist such as Equinix, Focal Communications Corporation, and Tel^{X} that maintain and operate NAPs of their own. Most such facilities aim to serve independent ISPs, but in some cases local competitive access providers also will be accommodated.

If wireless broadband operators want to confine themselves to offering nothing more than local high-speed access, none of this matters terribly much, but if they want to offer or support conferencing (particularly videoconferencing), streaming media services, Web hosting services, IP storage, real-time interactive applications such as multiplayer gaming, and pure end-to-end IP telephony, then the nature of the Internet connection becomes crucially important.

The Internet, as I have indicated previously, was conceived as a best-effort delivery network where the predictability of a connection was sacrificed in the interest of overall robustness and redundancy. The Transmission Control Protocol/Internet Protocol (TCP/IP) suite was never designed to support low-latency, constant bit rate transmissions, and although some of the older ancillary protocols such as UDP provided some support for delay-sensitive applications, the Internet never matched the ATM networks set up by the large telcos in terms of QoS.

Because of this basic deficiency, the major router manufacturers (Cisco Systems and Juniper Networks) have strongly supported standards (principally MPLS, RSVP, and DiffServ) that would enable routers to vie with ATM switches in supporting full QoS across a wide range of applications. Indeed, some would contend that current MPLS router/switches actually do a better job of supporting differentiated services than do legacy ATM switches.

Today most of the large carrier-class IP routers on the market are MPLS enabled and can support QoS for multimedia and real-time interactive applications, but that does not mean that a high-speed, high-fidelity multimedia transmission sent out over the public network is necessarily going to arrive intact at its destination. QoS enforced through the implementation of MPLS will result in more efficient use of bandwidth than would be the case carrying IP over ATM, but it still involves reservation of bandwidth, which means that bandwidth cannot be made available to anyone else until the transmission is finished. Time-sensitive transmissions simply make greater demands on capacity, and capacity always comes at a price particularly in the metro and the last mile. Long-distance carriers can choose not to meet those demands.

MPLS functionality can be partially or completely disabled in a router, and since ISPs and long-distance service providers are loathe to tax capacity to support QoS, delay-sensitive traffic is apt to encounter such a disabled router in its journey across the Internet to its final destination. If that occurs, it does not matter how stringent QoS provisions were in the path already traversed. The transmission essentially returns to best effort.

Some ISPs and long-distance carriers are willing to negotiate service-level contracts with local broadband access providers that will provide end-to-end QoS over an MPLS backbone. If broadband wireless operators want to launch transmissions outside their own network that require more than best effort, then such agreements are essential.

The Central Office As War Room

This chapter's intent has been to induce the potential network operator to view the central office and central network hub as a dynamic and evolving entity whose evolution will reflect a changing service model. Any network operator who survives and succeeds is going to have to retrofit the central office on a fairly frequent basis, and these operators should obtain facilities that will easily accommodate growth and change. While the basic network elements are not apt to change significantly within the foreseeable future, the operator will probably require greater and greater computing power and storage capacity to run the new applications that will surely emerge in the marketplace. I cannot emphasize enough that broadband services no longer constitute a utility business. Rather than providing some fixed offering, the network service provider is increasingly called upon to provide solutions to individual communications problems. In a real sense, the ability to innovate is becoming a core capability for the last-mile service provider.

CHAPTER 7

Service Deployments over Public Wireless MANs

This chapter focuses on service deployment; specifically, it covers the deployment of *value-added services*, which go beyond mere high-speed access and which serve to differentiate a service provider from its competitors. In this chapter, I go into considerable detail regarding standards and protocols and elucidate comprehensive procedures for implementing each type of service under consideration. As with so much else in this book, what follows is necessarily provisional simply because the standards, protocols, and physical platforms embodying them are in a state of flux and are likely to remain so within the foreseeable future.

Introducing the Pure Packet Services Model

As indicated many times previously, the aim of this book is to focus on packet services, specifically those using Internet protocols. This thrust is for a number of reasons.

First among them is the growing consensus among industry analysts that packet networks will come to dominate telephony in the years to come, not only in wireless networks but in wireline as well. A transition from asynchronous transfer mode (ATM) to Internet Protocol (IP) backbones is rapidly occurring in the long haul, and IP voice and packet data services are beginning to displace circuit voice and traditional Time Division Multiplexing (TDM) synchronous optical network (SONET) data connections. Even mobile networks are beginning a slow but inexorable migration to pure packet.

The second reason has to do with network efficiency. Packet networks are much more efficient than their circuit-based counterparts and will grow more efficient in the future. And because the wireless broadband operator is almost always challenged in respect to available bandwidth, network efficiency is no small matter.

Furthermore, because wireline incumbents have been notably slow in migrating over to packet services, particularly IP voice and Ethernet and IP business services, the new wireless broadband service provider utilizing the latest high-efficiency packet switches and routers possesses the means of neutralizing to a considerable extent the incumbent's advantage in overall bandwidth. By using and reusing available bandwidth with the utmost efficiency, the insurgent wireless service provider can frequently offer higher speed at lower cost than the incumbent—at good profit margins, to boot.

The Packet Model and Converged Services

The service model I propose is one of converged services. In other words, one network provides a multitude of diverse services. Traditionally, public networks have tended to be designed around a specific individual service such as telephony, video, text data, and so on, and the notion that one physical infrastructure could serve diverse needs is fairly recent. That this model will prevail in the future seems altogether likely, however, because growing numbers of broadband access providers have already begun to move toward it. Cable television operators now routinely offer telephone service and data as well as video programming, and telephone incumbents have begun to experiment with video over digital subscriber line (DSL). In this respect, broadband wireless is actually laggard, with most service providers still clinging to the pure access model.

The remainder of this chapter discusses in detail how the various services are delivered over a converged network, specifically, a wireless converged network, though it must be said that in the case of certain services the delivery methods are still imperfectly realized. At the time of this writing, the fully converged network remains more an ideal toward which service providers are striving than a fully mature entity. However—and this cannot be emphasized too strongly—convergence is already well under way in a growing number of broadband local access networks.

The key to convergence lies within the implementation of certain changes within layers 2 and 3 of the network, and within the last several years a multitude of protocols have been developed that will supposedly permit a packet network to emulate fully the characteristics of a deterministic network (in other words, a circuit network, so-called because the prior allocation of bandwidth that is the primary characteristic of a circuit network makes traffic flow highly predictable). The aim of such protocols is to preserve in large measure the bandwidth efficiency of best-effort packet networks while controlling traffic in such a manner that transmissions that require bandwidth on tap, so to speak, or that are intolerant of delays and timing errors will arrive intact at their respective destinations. This, unfortunately, is easier said than done, because it turns out to be extremely difficult to formulate rules that will meet simultaneously all the varying service requirements for diverse traffic flows while maximizing efficiency.

In essence, in such a converged network the router is continually shuffling packets from various flows, holding them in buffers known as *queues*, and releasing them from the queues to take their place in a single stream representing the converged IP pipe. In some cases, the network controller will also be strategically dropping packets to maintain desired flow characteristics. No protocol yet has been devised that is completely successful in emulating the predictability in service levels associated with circuit transports while maintaining high efficiency, particularly in the face of increasing network congestion. Some protocols perform better during certain traffic conditions than others, but all misbehave under certain circumstances. Obviously, if the router had an infinite amount of time to scrutinize traffic conditions from moment to moment, it could devise compromises through brute-force iterative approximations that would promote equity while satisfying bandwidth constraints, but the router has almost no time, so a fairly rigid, rules-based approach is required. Because no rules yet exist that conduce to ideal network performance, the network operator has a dilemma.

Currently, *best-effort* network traffic predominates within IP networks including the public Internet, so the problem is not yet acute. However, already IP voice telephony is beginning to expand rapidly, and with it the demands on the network for increased determinacy and predictable performance are increasing. A less rapid but still discernible expansion is occurring in the conferencing area (specifically, videoconferencing), and this, too, is demanding consistency in performance far beyond anything expected in the past. If, as I anticipate, increasing amounts of multimedia entertainment content are delivered over IP networks in the future, then nothing less than a thoroughgoing transformation will have to occur, and a high degree of network determinacy must become the norm, not the exception.

Wireless broadband operators, of course, must concern themselves with the here and now, and in the sections that follow I discuss what is feasible both in terms of the capabilities of the existing technologies and in terms of what is in fact supported over existing telecommunications backbones. In many cases, local access providers will not be able to provide service-level guarantees that are as stringent as they and their customers would like, but it is certainly possible today to support many time-sensitive applications over a last-mile broadband access network, including a wireless network.

But before I discuss the more demanding services in respect to quality of service (QoS), I first cover best-effort packet services, which, at least for the near term, will be the predominant service offerings by broadband service providers, both wireless and wireline.

Introducing Basic Access and Best-Effort Delivery

The majority of subscribers for broadband services today are apt to request nothing more than basic high-speed access. Such access will afford them a connection via the broadband wireless network to an Internet point of presence that will then take them onto the public Internet. Wireless broadband operators either can serve as independent Internet service providers (ISPs) and manage the connection to the point of presence or can off-load Internet traffic at a tandem switch in a telco central office and allow a third-party ISP to manage the connection to the Internet via a large router. They can also do both, offering ISP services to those subscribers who want them and simply providing an Internet connection to others.

In offering ISP services, broadband wireless operators will, at the least, need to maintain a server to cache e-mail and another to handle chat and instant messaging. They may also want to offer various value adds such as news reporting services, financial reports, and discussion groups. They may even want to sell advertising on the network's home page.

Space does not permit any detailed discussion of the logistics of launching and operating an ISP, and, in the case of wireless broadband service providers, any offerings in this area will be secondary to the speed of the connection in attracting subscribers. Generally, for the kind of business customer who will be the target market for broadband wireless service, consumer-oriented Web content pushed to the home page will not be much of an attraction.

The broadband wireless operator may also choose to function as a specialized ISP, offering such services as Web hosting and expedited content delivery, but this is a distinctly different business than simply providing access and is likely to involve investments and time commitments that are fully equal to those associated with establishing a metro area wireless access network. Since the focus of this book is on the latter, nothing more will be said concerning this option.

Here an extended definition of best effort and basic access is perhaps in order: Basic access provides use of the public Internet as a pipe for connecting to the Web and coincidentally to any remote office or business partners who are accessible through it—and little more. The main parameter in such basic access services is sheer throughput.

Throughput may be stated in terms of either maximum or burstable bandwidth or a guaranteed minimum. Unless a stated minimum rate is quoted, the service must be considered best effort with no guarantees regarding any performance metric.

It should be understood that in an IP network, without QoS provisions throughput will vary according to the numbers of users occupying a channel, and the perceived speed of the network will depend on the degree to which it is oversubscribed. Therefore, a stated maximum will not be meaningful in terms of the user experience. Nevertheless, stated maximums are frequently advertised by broadband service providers rather than guaranteed minimums, with cable data services sinning the most grievously in this regard.

In any case, the speed of the local access technology, whether it is wireless or wireline, is but one factor in the speed with which transmissions or transactions execute over the Internet. Regardless of the capacity of the access pipe, the user will experience delay and congestion if the routing paths by which Web sites are accessed are themselves congested or ill chosen, and the slowest segment of the network end to end will determine the speed of the transmission. Thus, if a user were to enjoy a 700Mbps connection to a central office and that in turn connected to an Internet point of presence via a DS3 SONET connection operating at a mere 45Mbps, then the maximum throughput to the user would be 45Mbps and no more. The actual throughput from a remote site over the public Internet would almost certainly be much slower and would ultimately set the speed of the transmission end to end.

In most cases, providers of local high-speed access specify only the speed of the link back to the central office, not to the Internet point of presence and not to some remote location that would involve a route across the public Internet. In the case of small businesses, a specified rate to the central office is often all that is demanded, but large enterprise users frequently want a good deal more out of their high-speed data services. At the least they are likely to want secure virtual private networks (VPNs), affording them reliable links to remote offices and often to business partners as well, and in many cases they are likely to require a good deal of determinacy in the network to enable them to do videoconferencing and/or the transmission of multimedia instructional materials in real time or near real time. Some will even want to perform such bandwidth-intensive functions as collaborative scientific computing. For time-sensitive applications such as videoconferencing, best effort will not suffice, and the subscriber will want guarantees regarding minimum throughput rates, latency, jitter, packet loss, and so forth—topics I discuss in a succeeding section.

To provide best-effort services over an 802.16 network, the network operator needs to assign the user only an IP address, whether dynamically or permanently, and allow the subscriber terminal to poll the network in the usual manner to gain access. All of this will normally occur transparently for the subscriber, and the polling process should occur more or less instantaneously except in cases of severe network congestion. Broadband wireless, like cable data, is essentially an always-on access technology with no logon procedure required, and thus the subscriber may expect nearly immediate response to keyboard prompts for online information.

In respect to network administration, the only real concerns for the network operator regarding basic access services are protecting customer privacy and guarding against attacks and unauthorized entry and utilization of network resources. The 802.16 standard contains security specifications within the standard pertaining to encryption, so subscriber traffic can be quite well secured with 802.16-compliant equipment if the standard encryption is utilized. Preventing unauthorized access to network resources is somewhat more involved, however, and may entail a number of different approaches.

As indicated previously, user access is generally regulated through an authentication server utilizing secret information stored in the subscriber terminal. Such information will involve layers of encryption such that eavesdropping alone will not permit intruders to uncover authentication codes. As 802.16 equipment proliferates and 802.16 radio modem cards come to be offered as accessories or even standard equipment for laptop computers, the importance of proper authentication procedures will grow. Already authentication has assumed major importance in 802.11 wireless LANs (WLANs), many of which are pathetically easy to breach, and it may be assumed that 802.16 service providers will learn from experiences of WLAN operators and build strict security into their networks from the onset.

The other aspect of controlling access and protecting subscribers involves the firewall that authenticates on the packet rather than the session level and that continually monitors and filters traffic coming into the network. In many cases subscribers will want to maintain and administer their own firewalls, particularly if they are business customers. However, in broadband services aimed at residential customers, particularly in the cable data business, the service provider may well assume responsibility for firewall functionality in the central office. Some edge routers such as the Redback Networks products have built-in firewall functionality, but in other cases a separate device will be required. Incidentally, software firewalls installed in servers devoted to other uses are generally unsatisfactory in a public network setting because of the claims on computing resources and their tendency to slow the entire network. Chapter 8 covers firewalls and other security considerations at greater length.

Moving Beyond Basic Access: VPNs and LAN Extension

I have mentioned VPNs previously and have identified them as key service offerings for addressing the business community. Many enterprises, both large and small, choose to set up VPNs on their own without any service provider intervention, but the smallest businesses frequently lack the in-house expertise to manage a VPN, and, in such cases, the service provider can offer the provisioning of a VPN as a value-added service. Increasingly, broadband service providers are doing so, and the wireless broadband operator who hopes to compete successfully in today's access market must be prepared to meet this need.

Here perhaps a bit of history is in order: VPNs arose out of the understandable desire on the part of organizations with geographically distributed offices and operating centers to provide branches with full access to the centralized databases at the corporate headquarters and thus to avoid having to duplicate network resources at each location. Up until the late 1990s, wide area enterprise networks were generally connected through leased circuit connections such as T1s or DS3—inevitably an expensive proposition when multiple locations were involved, particularly where any-to-any connections were mandated. Alternately, ATM or frame relay virtual circuits could be used with nearly equal reliability and at substantially lower cost, though frame relay did not scale well when multiple remote locations were involved. Any of these legacy connectivity technologies provided a high degree of service reliability and consistency as well as a fair measure of inherent security, but the pricing of such services was such as to place them out of reach for many businesses.

VPNs, even in their developmental stage, took a number of forms. VPNs could be restricted solely to the branch locations of a single business or governmental agency, in which case they were known as *intranets*, but many enterprise managers chose to extend them to close business partners, particularly those with whom they had ongoing relationships and with whom they performed frequent transactions. If a trusted contractor were making daily claims on the resources of a company, it made sense for the host company to eliminate bureaucratic processes as much as possible and allow the contractor direct access to relevant databases. When such an outside party was involved, the VPN would then be termed an *extranet.*

As a VPN grew to encompass organizations outside the individual enterprise that had originally set it up, the problems in administering it grew as well, because at that point network administrators had to deal with two or more sets of internal IP addresses, some of which may overlap. They may also have to deal with various multicast groups, each of which would receive certain information intended for it alone. Worst of all, administrators, because of a flurry of IP standards activity, faced difficult decisions as to which standards-based approach to implementing VPNs would best meet the needs of their operations. Accordingly, at the time of this writing, VNP administration has become almost a discipline unto itself.

Any complete discussion of VPN setup and administration goes far beyond the scope of this book, but I will outline the basic approaches and indicate where they are apt to produce the most satisfactory results. The following sections consider the practical implications of VPN administration from the service provider perspective and in particular how to make a business out of supporting VPNs.

Types of VPN and the Role of Network Operator in Administering Them

Currently, a plethora of approaches to creating VPNs exists, and, because of ongoing vigorous activity by standards bodies, you are apt to see methodologies continue to proliferate, at least in the midterm.

Probably the most basic distinction among the several types has to do with the layer of the network stack in which VPN creation takes place. All variants may very generally be placed within three basic categories, though inevitably there are various degrees of hybridization.

The first category consists of subscriber-managed VPNs that are essentially applications. The second category is comprised of service provider implementations where layer 2 switching is utilized to segregate VPN traffic. Finally, the third consists of those service provider realizations where layer 3 routing defines the traffic flows.

Subscriber-Managed VPNs

The first category is not properly the province of the service provider at all. Here the subscriber simply leases an Internet connection from the service provider and takes full responsibility for setting up the *tunnels* (described shortly) over which VPN transmissions take place. Customer premises equipment assigns the VPN traffic to various tunnels, and the service operator has no role in traffic handling beyond maintaining an IP pipe to the customer. Essentially VPNs of this type cannot be considered services and thus bring little benefit to the service provider beyond perhaps some additional sales of bandwidth. Some service providers find this arrangement appealing since it makes no particular demands upon them, but, on the other hand, it does not allow the service provider to charge the customer extra for VPN services. In other words, it is not a profit center.

Subscriber-administered VPNs are quite common among large enterprise customers who generally have in their employ experienced LAN managers, and for some organizations these privately operated VPNs are entirely satisfactory. But where an enterprise manager must connect with widely scattered remote locations, and especially where specific amounts of bandwidth and stringent constraints on delay are demanded, then a private implementation is unlikely to prove wholly satisfactory. Therefore, wireless broadband operators should not rule out even the largest organizations as candidates for managed VPN services.

Switched VPNs

The second type of VPN, that defined by layer 2 switching, takes a number of forms based on both the protocols utilized and the setup methodology.

Such layer 2 VPNs can be established over pure circuit networks or over packet networks that employ switches rather than routers. Both approaches have their respective advantages and disadvantages.

Pure Circuit-Switched VPNs

I can dismiss in a few words pure circuit services. It is perfectly possible to create a VPN by concatenating T1 connections over distance, and such "nailed-up" VPNs were commonplace in the 1980s and early 1990s before telcos offered packet services. These TDM circuit-based VPNs provide a high degree of privacy and availability but are prohibitively expensive to set up on a large scale and, since the introduction of so-called fast packet services, have fallen into general disuse. In theory, a broadband wireless operator could set up a circuit-based VPN for a customer by allocating an entire channel to that customer and then arranging with a local and/or long-distance telco for the provision of a circuit connectivity through to the endpoint of the VPN, but this does not represent a service model with much potential for the future.

Switched Fast Packet VPNs

During the middle and late 1990s, VPNs utilizing frame relay and/or ATM connections—what have come to become known as *fast packet connections*, though ATM is not really packet based—became popular, and they are still present in the marketplace today, particularly deployments where frame relay traffic is encapsulated within an ATM overlay. Such VPNs are also known as *virtual private wire services (VPWSs)* because the connections themselves are treated as if they are hardwire circuit connections rather than virtual circuits—that is, fixed allocations of bandwidth set up by the ATM and frame relay protocols. In terms of ensuring privacy and high availability, there is nothing wrong with frame relay VPNs, but the speeds supported by frame relay service providers have been unimpressive, and the efficiency and robustness of frame relay networks are inferior to that of Multiprotocol Label Switching (MPLS)–enabled IP networks that offer similar benefits. Frame relay, because of a relative lack of redundancy and error correction mechanisms, is not well suited to the wireless realm and has seldom been employed in wireless broadband networks. However, as indicated in Chapter 1, the 802.16 standard does support ATM. In theory, an ATM-based VPN service offering is perfectly possible within a broadband wireless network, but it squanders precious bandwidth. Furthermore, in the case of remote locations outside of the wireless operator's own service area, it compels the operator to purchase expensive end-to-end ATM connections through the backbone from long-distance service providers.

Switched Packet VPNs

Packet streams can be either switched or routed, and the distinction is important, both generally and in regard to VPNs. Routers direct packet traffic one packet at a time and thus can send individual packets within the same transmission over diverse routes according to momentary fluctuations in network congestion and availability. Switches, on the other hand, establish a single physical path for the stream, and each packet follows that path regardless of network conditions. Thus, switches are relatively dumb compared to routers.

Discussions of circuit switching of packet traffic tend to give rise to some confusion, so I will attempt to clarify the issues.

In a pure circuit connection, such as that afforded by SONET, only one transmission will occupy any particular path. In a circuit-switched packet network, all packets in the same transmission will follow the same path as well, but they will not have that path for their exclusive use, as is the case with a switched circuit network. In a sense, the circuit-switched path approach may be seen as a kind of compromise between packet routing and pure circuit switching. Because multiple streams are contending for bandwidth within a single path, the efficiency of the network is higher than is case for circuit; however, network availability is lower. At the same time, the intelligence required of the network node is less than for a routed packet network.

Two particular switching protocols are employed within VPNs, Ethernet switching and MPLS, and sometimes both are used together. In both cases, and in routed VPNs as well, a technique known as *tunneling* is used to set up the VPN.

A tunnel is a means of transmitting data across a network from one point to another as though the two points were directly connected. The tunnel is established simply by providing the packets with an extra address header that is attached to the front of the packet. Any intermediate node in the network that has been enabled to support a VPN will pass the packet based on the forwarding instructions in the extra header without examining anything else in the packet. Incidentally, tunnels need not be restricted to one user. In certain tunneling protocols, several VPNs may be multiplexed within a single tunnel, though the privacy of the different users will be maintained.

In simple tunneling only the address is being used to ensure privacy. Each device attached to the network is being asked in effect not to read the packet and to pass it on. Thus, the privacy within such a virtual private network is conditional, and for subscribers whose transmissions are highly confidential—and that includes most subscribers to business-class data services—something more is needed, namely, encryption of the data stream.

Several methods exist for actually executing a packet-switched VPN across a network. Those methods fall into two subsets, one utilizing an Ethernet switch and the other an MPLS switch, both located at the service provider's central office.

The first of the switched Ethernet approaches you shall consider is a virtual private LAN service (VPLS). Here the enterprise LAN extends to the edge of the service provider network, and the provider network assumes the function of an Ethernet switch to connect all the individual remote locations into a single logical-switched Ethernet. Customer premise equipment performs no traffic direction function in regard to the VPN. The service provider handles all that.

Closely related to the VPLS is the IPLS (which stands for IP-only LAN-like service). Here IP rather than Ethernet traffic emanates from the subscriber premises, but that traffic is switched rather than routed by the service provider. Inasmuch as 802.16 is based on an IP interface between the subscriber and service provider terminals, IPLS is easier to implement than VPLS, but, for this type of VPN to be set up, subscribers must have routers of their own. Incidentally, IPLS solutions are not standards based as yet.

MPLS switching, which seems to represent the more current approach, provides network operators with a number of possibilities in regard to VPNs. They can utilize ATM or frame relay virtual circuits over MPLS-switched label paths, which is a fairly simple procedure but does not scale well. Alternatively, they can utilize the Martini draft extensions of MPLS that have not yet been fully standardized by the Internet Engineering Task Force (IETF) but are supported by many MPLS switches today. This approach does in fact scale well.

Yet another approach is the Kompella layer 2 VPN. This is simpler to administer for the network operator than Martini draft, and it is highly scalable.

Layer 3–Based VPNs

In terms of standards-based solutions, VPNs based on layer 3 are further advanced than their layer 2 counterparts. This being the case, equipment choices will be more varied and protocols will be more stable. Two such layer 3-based approaches are extant today: RFC2547 and virtual routing.

RFC2547, also known as Border Gateway Protocol (BGP) over MPLS, is probably the most commonly used service provider VPN technology today. (BGP is a frequently used routing protocol across the public Internet that has been long standardized.) Plain-vanilla BGP will not do, however, because of its inability to handle overlapping addresses, so a route distinguisher extension is required. In this scheme, a separate MPLS label will be associated with each VPN, and BGP will be used to advertise the label to each intermediate node encompassed by the route.

Virtual routing is somewhat different in that a separate routing announcement disseminates information on each routing path associated with each VPN, but in RFC2547 a single announcement is made. Virtual routing is more difficult to administer than RFC2547 because of its obviously much greater complexity and is less used in the industry. Its chief advantage is the extreme isolation that it provides for individual traffic streams and its extensive support for fine gradations of QoS as applied to many individual transmissions within the VPN.

Tunneling Protocols

Setting up and maintaining tunnels is separate from the routing and switching functions that direct information into those tunnels. As you have seen previously, tunneling generally involves adding a header to a packet that will define the tunnel, but this is not the case with all protocols. In the case of MPLS, a tunnel can also be defined with a switching label, avoiding the use of any ancillary protocol.

Those protocols used specifically for tunneling include IPsec, Layer 2 Tunneling Protocol (L2TP), IP in IP, and Generic Routing Encapsulation (GRE).

IPsec is basically an encryption protocol that segregates streams by applying separate encryption codes to them. It lacks flexibility compared with some of the other tunneling protocols but can be combined with most of them to redress that deficiency.

L2TP is designed to tunnel layer 2 information across a layer 3 network. It is often employed in switched VPNs.

IP over IP is little used today and can be passed over. GRE, which serves the same purpose (namely, to define IP tunnels over a pure IP network), is much more flexible and contains IP extensions that allow the network operator to make do with a relatively small number of IP addresses and to multiplex tunnels.

Applications Requiring Quality of Service

This section concerns value-added services that are relatively intolerant of variations in throughput rate and availability. For the most part, they involve multimedia such as voice or video, but they may also encompass real-time scientific computing, gaming, white boarding, voice-over business presentations, distance learning, financial reporting, and indeed any application that is both data intensive and dependent on an orderly and predictable succession of data.

Applications will definitely figure more prominently in the future, but the speed with which they will be adopted cannot be predicted. In the late 1980s and early 1990s, many industry analysts assumed that multimedia and "virtual reality" interactivity would typify network applications by the turn of the century, which, of course, was not the case. Web-based multimedia, such as it is, has consisted, for the most part, of short video clips of poor to middling quality, some real-time audio in the form of private conferencing or Internet radio, and an abundance of simple animation made possible through Java applets.

Why has Web-based multimedia been so slow in arriving? Answering that question conclusively would require a depth of analysis far beyond the scope of this book. What I can say is that such applications have yet to reveal compelling advantages over more traditional broadcast and circuit-based multimedia in any but a few areas. To cite just a couple of examples, Internet radio and television are qualitatively inferior to standard analog radio and television and have foundered badly on that account, and they are difficult to deliver over conventional domestic and portable entertainment systems. Furthermore, Internet radio is currently infeasible in the mobile environment where most radio listening occurs. Web-based multimedia simply cannot compete effectively with the older circuit-based content delivery systems, and they will not be able to do so until the underlying technology improves significantly. Indeed, only in the areas of conferencing, distance learning, and gaming do the interactive dimension afforded by two-way broadband confer a decisive advantage, and it is precisely these applications that show the most promise of attracting substantial numbers of subscribers in the midterm.

QoS is a rather imprecise term, particularly when applied to packet networks. Preeminently, QoS is an end result, one that can be achieved through a number of different transport technologies.

Generally, the term refers to conditions within the network that will support the delivery of time-sensitive or low-redundancy services with minimal perceptible degradation. More particularly, QoS encompasses the following: control of throughput rate, specifically the minimum rate; control of overall delay or latency; control of packet-to-packet delay, known as *jitter*; and control of packet loss or bit error rate. Some would also include availability and the likelihood of call or session blockage as well as subjective impressions of presentation quality—the latter constituting an especially important issue when the network operator is utilizing data compression technologies.

I will present each of these attributes in turn in reference to the various services that a broadband wireless operator is apt to offer.

Throughput rate refers to the number of bits per second in a digital network. It is a fairly important metric in the delivery of most services, but it is vitally important in the delivery of services such as high-resolution video that inherently require large amounts of data to be delivered more or less continuously.

Latency or overall delay tends to be particularly important in interactive applications such as voice telephony, conferencing, gaming, and certain instructional applications where individuals are being guided through a process by a human instructor or by information summoned from a database. Latency is also important in the transmission of protocols such as Transmission Control Protocol/Internet Protocol (TCP/IP) that depend on frequent acknowledgments.

It is perfectly possible to have abundant bandwidth and throughput and, at the same time, to experience significant latency. An excellent example is high-speed geosynchronous satellite services where the round-trip delays engendered by the two 24,000-mile journeys from uplink to downlink for each packet result in overall delays that are a substantial fraction of a second and are readily discernible in telephone conversations. Significant latency may also arise from a packet having to pass through a large number of switches or routers within a terrestrial network even while the throughput remains quite high.

Jitter is peculiar to routed packet network and arises when the time required for a packet to transit the network varies from moment to moment. Jitter is particularly injurious to real-time interactive multimedia such as ordinary voice telephony. Jitter may be traded for delay by storing packets in a buffer and then releasing them with equal time intervals between them.

Packet loss or *error rate* represents the failure of the network to deliver all the packets transmitted to the recipient. Normally some packet loss occurs in any network, but in wireless networks the incidence of packet loss tends to be higher than in wireline networks because of fluctuating background interference levels, sudden fades because of multipath, and variable attenuation of the signal with changing weather conditions. Buffers will also drop packets when the buffers become completely full, and thus packet loss will be the normal attendant of severe jitter.

Many packet protocols contain provisions for checking data integrity at the receiving end and for retransmission in the event of severe packet losses. The problem with that strategy, however, particularly in a radio network, is that the amount of radiated energy and claims upon bandwidth necessarily increase in the presence of retransmissions, and the retransmitted packets are delayed in relationship to those not requiring retransmissions, a state of affairs that will greatly increase delay, jitter, or both.

Packet loss is problematic for any data transmission but, interestingly, generally has less effect on multimedia than on textual data. Humans can resolve images and audio transmissions in the presence of significant reductions in information density or in the presence of false data simply because sounds and images contain so much redundancy. But in the case of textual data, the loss or distortion of individual symbols can be critical to understanding the text. On the other hand, delay and jitter pose few problems for text transmissions, so the sender can retransmit fairly freely to minimize error rate without worrying much about the order in which packets arrive or their time of arrival.

Availability I have already dealt with in preceding sections, and I will not revisit the subject here. Availability can apply equally well to best-effort services and is not peculiar to services meeting the more general conditions designated by QoS. Certainly availability carries connotations of quality, but the term itself refers to no specific attributes of the service, which is generally what one has in mind in discussions of QoS.

Subjective assessments of quality, or additional dimension of QoS, generally apply to audio or video presentations. Here the user will be alert to *artifacts*, which are departures from fidelity because of data loss or the imperfect operation of compression algorithms. Such departures do not lend themselves to easy quantification or the assignment of one-dimensional metrics, and the sensitivity of individual users to such aberrations may vary considerably. To one individual, MP3 audio compression at an 11-to-1 ratio may be perfectly transparent while to another may be unacceptable in terms of sound quality, and that is why such assessments are deemed subjective. Normally, service-level agreements, considered at length in the following section, avoid subjective considerations and focus on the readily quantifiable.

Marketing QoS Through Service-Level Agreements

QoS, regardless of how it is achieved, is inherently more expensive for the service provider than is best-effort data delivery, given an equal allocation of bandwidth. This being the case, service providers will have the expectation of commanding a premium for QoS in their networks.

In a packet network, QoS will vary from application to application and from customer to customer. Normally, part of the available bandwidth will be given over to best-effort services and part to services demanding QoS. Incidentally, such apportionment is possible only within a packet network. A pure circuit network such as a SONET ring can offer bandwidth to subscribers only in fixed increments, each with the inherent QoS that comes with wholly predictable bandwidth assignments.

Some types of services such as voice telephony and regular video programming provoke in the subscriber an expectation of high presentation quality simply because such services were delivered exclusively over circuit networks in the past and typically exhibited excellent fidelity. For this reason no one is likely to opt for grainy, pixilated video presentation even at a reduced fee or for robotic voice quality and intermittent phone service. In such cases the network operator simply cannot sell on the basis of a guaranteed minimum level of service quality; rather, the service has to be perceived as good to excellent.

Nevertheless, for many business services, including conferencing and LAN connectivity, the operator can offer stratified services with differential pricing. For example, videoconferencing may be offered at a medium throughput of 400Kbps or at a high throughput exceeding 700Kbps, with the latter commanding premium pricing. The subscriber can expect near broadcast quality at the higher figure, and some video artifacts at the lower figure, but because 700Kbps pipes are quite expensive over long distances, the subscriber may opt for some degradation in picture quality to stay within a budget. Because the previously prevalent Integrated Services Digital Network (ISDN) videoconferencing was so compromised in fidelity, it set expectations for the service at a fairly low level—unlike the case with voice telephony where the original analog transmissions established a standard of excellence that has still not entirely been met by packet services—and thus many users are perfectly satisfied with compromised quality if the price is right.

Where tiered services are offered, a service-level agreement will establish the performance that the subscriber may expect. Normally, performance will be established in terms of minimums that must be met by the service provider rather than the best performance possible over a lightly loaded network. In most cases, the subscriber will be entitled to a discount on the monthly fees when service levels are not maintained according to the agreement.

Achieving QoS in Broadband Wireless Packet Networks

As indicated many times previously in this text, the core packet standards in common use today (namely, TCP/IP and Ethernet) were both originally intended for transmitting best-effort traffic only, not traffic requiring stringent QoS. As a consequence, equipment manufacturers and standards bodies have made a great deal of effort in recent years to enable packet networks to handle QoS through both ancillary protocols and revisions of the core protocols. The 802.16 standard has a number of provisions for ensuring QoS, and it permits the use of a wide range of additional protocols for further ensuring that quality levels will be met.

In addition to the industry standards supporting QoS, there are a number of other strategies for achieving it, including overprovisioning of bandwidth, data compression, and special routing protocols combined, often as not, with Internet diagnostics and route analysis software. I will cover each of these later in this chapter.

It should be clearly understood that establishing QoS is much easier within the confines of a metropolitan network than it is across great distances where the transmission must traverse several backbone networks on the way to its final destination. If, for instance, network operators want to send video material cached at the central office to a local subscriber, they can simply provision enough bandwidth to support the transmission and can rest assured that the quality will be acceptable, barring some catastrophic interruption of the link. If, on the other hand, operators are hosting a videoconference that spans the continent, they are at the mercy of those parties controlling intervening network segments. They can contract with a long-distance service provider that will agree to support a service-level agreement, but they cannot personally allocate bandwidth across the network to ensure that all relevant performance parameters will be met.

The QoS provisions within the 802.16 standard are generally sufficient to provide for good QoS on applications running entirely within the network, including voice- and broadcast-quality video. They are nearly useless, however, in the long haul, and the network operator must utilize additional protocols and, in most cases, forge agreements with other service providers controlling intervening data paths and network nodes.

802.16 Provisions for QoS

At the physical layer, 802.16 is actually a circuit-based protocol. It allows for channel divisions based upon frequency divisions or time slots, with time slots expected to predominate in the radios conforming to the standard.

Recurrent time slots separated by intervals of other time slots can be reserved entirely for certain subscribers or shared among numbers of subscribers. When sequences of slots are reserved, they function analogously to the time slots or temporal channels in a SONET network. Such strict reservation of the bandwidth in one set of slots provides inherent QoS for a subscriber, albeit at the cost of efficiency, as is the case with all session- or circuit-based approaches. Depending on the types of services the network operator has settled on, some sequences of slots may be used to provide circuit service while others will be allocated to best effort. However, it should be understood that circuit-like determinacy can be delivered by means other than reserving time slots for the exclusive use of a single party.

The 802.16 standard also provides inherent support for ATM, covered next, and IP 6, which provides for flow labeling somewhat in the manner of MPLS, also covered next. In and of itself, IP 6 does not provide extensive support for QoS, however, and cannot be considered adequate absent the use of additional protocols along with it.

Major Networking Standards for Supporting IP QoS

To a considerable extent, the standards for achieving QoS with IP transmissions are complementary in the sense that none represents a complete solution. Frequently two or more protocols are used together.

IP over ATM

The 802.16 standard fully supports ATM, which itself offers excellent support for QoS on quite a number of applications. ATM is also supported almost universally by long-distance service providers, and unlike the case with MPLS, QoS functionality will be fully enabled end to end across the network when transmissions are directed through ATM core switches. ATM is, however, less efficient and less flexible than MPLS, and ATM services have not been competitively priced in the past. I consider ATM to be a legacy technology that is not ideally suited to the new service provider striving to compete on the basis of value with entrenched incumbents.

Commonly Employed Ancillary IP Protocols for QoS

From the middle 1990s onward, when the ascendancy of IP was becoming obvious, certain IP advocates sought to redress its equally obvious shortcomings in respect to determinacy by creating supplementary protocols. The fact that so many of these have been created would seem to argue the individual inadequacy of any one of them, and in fact they are frequently used in multiples. The protocols I will consider include RSVP, IntServ, DiffServ, IP Precedence, RTP, RTSP, and MPLS, among others.

Most such protocols perform one of three functions. First, they request a certain allocation of bandwidth from the network deemed necessary for the execution of the application they are intended to support, and then they attempt to establish separate streams for individual applications where certain minimum network performance requirements will at all times be upheld. Thus, real-time video may go into one stream, toll-quality voice into another, and so on. Second, they append extra information to the packet in the form of an additional header or label. This assigns the packet to the correct stream and also requests special treatment by any network element the packet subsequently encounters; in effect it is a service classification label. Finally, they invoke specific mechanisms at the router for dealing with congestion and delay where different instructions for emptying queues, dropping packets, and retaining packets within queues over time are applied to different streams.

It should be understood that the protocols themselves form only one aspect of a unified approach to QoS and that the use of the protocols as Band-Aids on an underengineered network is not going to produce satisfactory results. The protocols enable bandwidth to be used efficiently in supporting various services but they do not create bandwidth where it does not exist. In a sense compression can do that, but that is another issue for another section.

RSVP

Resource Reservation Protocol (RSVP) is a protocol for requesting bandwidth from routers along the path that a transmission will take. It does not guarantee that adequate resources will be provided, however, because one or more of the routers may themselves either lack adequate resources or may not have the protocol enabled. RSVP itself is not a routing protocol, and it makes no determination regarding the path that packet will take. Instead, it is a *flow-based protocol* because it merely signals the requirements for various traffic flows. Incidentally, RSVP requests must reach every router or switch in every conceivable signal path the packet may take, so it tends not to scale well over distance. RSVP does not employ extra headers or labels, but it frames its requests for bandwidth within the normal IP header in the form of priority bits.

IntServ, DiffServ, and IP Precedence

IntServ stands for *integrated services* and was a protocol developed by the IETF in the 1990s. It is little used today, and I will pass over it here.

DiffServ, which is widely used today, performs a somewhat different function than RSVP; it provides a classification scheme for various streams rather than requesting bandwidth along a data path. DiffServe, as well as RSVP, requires that priority bits be assigned to the packet header to establish the priority of that packet in terms of the queue it will occupy and the order in which it will be transmitted relative to packets of other priorities occupying other queues. Such queuing will take place at switches rather than routers, whether MPLS switches (increasingly the case today), ATM switches, or metro Ethernet switches.

As is the case with RSVP, all relevant network elements must be enabled for DiffServ, or QoS will not be maintained across the network.

IP Precedence fulfills a similar function to DiffServ, but the latter is much more widely deployed today. IP Precedence, unfortunately, is not consistently implemented from one vendor to another and thus is of limited usefulness in real-world networks.

RTP and RTSP

Real Time Protocol (RTP) and Real Time Streaming Protocol (RTSP) are more specialized than the other protocols covered thus far, being designed primarily to support audio and video streaming. RTP, which has somewhat broader application than RTSP, is used for a number of real-time applications, including but not limited to audio and video. RTSP is used mainly in multicast or media-on-demand applications and in IP telephony as well. RTP is a transport protocol, and RTSP, as the name implies, is as streaming protocol used for transient single-session multimedia presentations rather than the transfer of storable multimedia files. RTSP is designed specifically to work in a client-server environment.

MPLS

MPLS is a switching protocol that supports QoS by permitting the setup of label-switched paths for traffic of similar characteristics. It is not, as I have seen, intended primarily for QoS but rather for effective network management.

Queuing Techniques That Support QoS

In addition to the previous protocols, which primarily involve signaling, a number of protocols promote QoS by managing the relationship between and among queues assigned to different classes of traffic. These include strict priority queuing, round-robin queuing, fair queuing, weighted fair queuing, adaptive scheduling, and so on. The first requires the complete emptying of the highest-priority queue before the switch proceeds to that which is next in priority, and the others strive to provide some degree of equity to lower-priority streams. No such protocol can be said to be perfect, and network operators will have to determine which will best satisfy their customers in aggregate. Normally these queuing protocols will not be the province of the broadband wireless access provider but of the owner of backbone fiber and/or Internet access points.

Furthermore, several protocols exist for managing connections end to end across the backbone, including Resource Allocation Protocol (RAP) and the aforementioned RSVP. Although the signals may originate in the metro, the actual resource allocation will take place on the backbone.

Making Sense of QoS Protocols

QoS protocols, as previously mentioned, are often used in multiples. RSVP and DiffServ, for instance, are often used in conjunction with MPLS to support a wide variety of applications. Other protocols may be used more selectively; for example, RSTP would be used for streaming video but not for voice telephony. Given the proliferation of protocols and the complexity of their interactions, engineering an end-to-end IP network for QoS is not an easy task, and it is in fact nearly impossible where protocols are not carried consistently across the network. Unfortunately, configuring intervening switches and routers is generally out of the hands of the local access provider.

With content demanding very stringent QoS such as high-definition video or Super Audio Compact Disc (SACD)–quality audio, delivery over a multimode network may simply be infeasible, and if the source is geographically remote, the local access provider may choose to utilize a satellite service to transmit the content in one hop to the local network where it can be cached on a server until needed. It can then be transmitted over a short distance where QoS is entirely under the access provider's control.

Other Methods for Supporting QoS

I should not leave the subject of QoS without mentioning a long-standing argument as to the need of special protocols for supporting it over packet networks. Some in fact would contend that QoS can be achieved through other means and that special reservation and prioritization protocols are not the answer because of the difficulty of implementing them across heterogeneous networks.

The simplest technique for getting around reservation requirements is to undersubscribe available bandwidth. If everyone on a network enjoys, say, a median throughput rate of 50Mbps, QoS is likely to be satisfactory for almost any application absent any additional provisions for ensuring QoS. And indeed some in the industry have suggested that as network throughputs inch ever upward, the need for special QoS protocols will disappear over time.

Given enough bandwidth, surely such representations are correct, but for now bandwidth is still expensive, particularly in metropolitan networks, and with some services such as lifeline telephony or backup storage, the user cannot tolerate momentary suspensions of service, however infrequent. For such services a reservation of bandwidth simply has to be in place against every contingency.

Another way to attempt to get around the need for reservation is to create a sort of artificial bandwidth by means of data compression technologies. Compression algorithms reduce the bandwidth required for any given application, often dramatically, and 50-to-1 reductions in the amounts of data required to transmit a signal are commonplace today. In practical terms this may approximate a 50-fold increase in bandwidth, though in reality matters are not so simple. Compression almost always introduces appreciable latency in a transmission, which puts one in a position of trading capacity off against presentation quality. Moreover, heavily compressed signals are highly intolerant of data loss because they have already been robbed of the redundancy to compensate for it. Compression buys the network operator something, but, as the saying goes, there is no free lunch.

Compression is a vast subject, much too vast to treat with any thoroughness here, but most compression techniques fall into a few basic categories. The simplest seek to remove redundant information in a signal and only register changes from one sample to the next. More sophisticated techniques use what are known as *codecs*, which are idealized models appropriate to a given type of information such as representations of the human voice or a physical depiction of the human form. The information that is to be conveyed is compared to the model, and only what is unique to the individual message is transmitted. Still other compression techniques are based on underlying regularities in the forms of the entities depicted in the data or on *perceptual codecs*, where features that ordinarily would go unnoticed by a human observer will not be transmitted. Whatever the individual compression technique, all such techniques fit into two overarching categories: lossless compression and lossy compression. In the former, the original message can be reconstructed perfectly even though not all the information in it is actually transmitted, and in lossless transmission a recognizable facsimile is created at the receive end that does not contain all of the original information.

The widely used MPEG compression schemes for audio and video are good examples of lossy compression depending upon perceptual codecs. They produce perfectly recognizable sounds and images, but, depending on the compression ratio (that is, the ratio of data removed to data preserved), they may degrade image and sound quality. The compression techniques in common use for audio and video today have grown very sophisticated, and at moderate compression ratios (say, five to one), they are well nigh undetectable by most viewers and listeners. And, inasmuch as this is true, they leverage existing bandwidth effectively. A telephone call that would require 64Kbps can be made with 8 or even 4 kilobits with little loss of fidelity. A high-definition television signal with several times the information per frame as a traditional National Television Standard Committee (NTSC) analog signal can yet be squeezed into a 6MHz NTSC channel while yielding a much more detailed image.

Compression is built into many Web tools used today such as real-time video and audio players and is not generally used as an overall strategy on the part of the network operator, though numerous product offerings are available for that purpose. The problem with using them is that they have to be enabled in every customer terminal that is to receive a compressed signal, so the logistics of using compression tends to be challenging. Nevertheless, I expect compression to assume greater importance over time.

The routing protocol used end to end will also have a considerable bearing on latency and jitter and hence on QoS. Again, this will primarily concern the operator or operators of the backbone, not the metro operator. Frequently, a message will pass through more than one long-haul network in traversing a great distance. Unfortunately, since many long-haul operators jealously guard routing information within their own networks and do not make it available to other network operators, a router in one network attempting to plot a route across intervening networks to a final destination may simply lack the information to make optimal choices with predictably unfortunate consequences in regard to QoS. To some extent service-level agreements between and among long-haul providers can obviate such difficulties, but often the party seeking to maintain QoS by this means will have to pay dearly to do so.

Finally, I should mention the methods for supporting QoS that use proprietary route analysis techniques but do not depend on special arrangements with long-haul service providers.

A specialized ISP called InterNAP claims to support QoS over the public Internet by such means if the customer utilizes an InterNAP facility to gain access to the backbone.

Where QoS Matters Most

While QoS is vitally important in any IP multimedia application, only two such applications have found much of a market to date. Those two are IP telephony and conferencing. Fortuitously, considerable overlap exists between the tools, both hardware and software, required for either application. In the future, perhaps in the not-too-distant future, other applications demanding stringent QoS will become routine, and the QoS required to enable top-flight videoconferencing, an application that is already in some demand, should suffice for nearly anything else.

Enabling IP Telephony

Because telephony is the value add that, apart from VPNs, appears to be most salable today, and because the technology behind it is quite complex, I am devoting an entire section to it.

IP telephony properly refers to a whole category of technologies, some complementary and some mutually exclusive. Simply stated, there are a number of different ways to do IP telephony. The end result may appear to be a single application to the end user, but the underlying mechanisms are many and diverse.

Any discussion of IP telephony should perhaps begin with some consideration of the rationale for doing it in the first place, and that largely has to do with using fewer network resources per call and building and operating a network for much less than would be the case with a circuit voice implementation. IP telephony network elements cost a fraction as much as traditional circuit switches with a similar port count and also require much less maintenance and manual configuration to operate. An IP telephone service can be operated with fewer staff members than a traditional circuit service—though how many fewer is a matter of debate—which is why few new telephone service providers are buying class 5 switches anymore. Furthermore, since the bandwidth required per call is much less for IP services, the network operator can use available bandwidth for provisioning higher-value data services or simply build less capacity into the network in the first place. In a wireless broadband network, where augmenting capacity generally means increasing the number of costly base stations, anything that increases network efficiency is a godsend, and thus Voice-over IP (VoIP) is incontestably preferable to circuit.

All of this being the case, one is moved to ask why IP networks are still relatively little used in transmitting telephone calls. Why is it that more than 90 percent of residential calls made in the United States are still transmitted over inefficient circuit networks?

This is for several reasons. First, by most estimates, IP telephone equipment has only recently reached the level of maturity where many carriers would want to rely on it. Second, IP backbones that could support the QoS necessary for IP voice have not been pervasive until now. Third, the price of circuit services has continued to decline, resulting in singular lack of urgency among subscribers to migrate to IP voice. Finally, the services themselves may be viewed as a somewhat inelegant overlay upon a network never intended to support them. Moreover, they have involved a multitude of palliatives and makeshifts that even today do not offer the full functionality of the mature PSTN. It does not help matters that these makeshifts and palliatives do not represent any unified approach to the problem.

Another issue has to do with the value of IP voice services to the end user. Because voice and data can be sent in parallel within the same transmission, IP voice holds the promise of delivering services that are either infeasible or prohibitively expensive with ordinary circuit voice. One can envision a number of interactive multimedia applications such as collaborative design sessions involving engineer, architects, or industrial designers where images are manipulated while changes are discussed or perhaps online sales presentations where a sales agent walks a customer through a video catalog and amplifies on the products. Images could even be controlled by voice prompts, raising the level of interactivity beyond anything that could be practically executed with circuit transmissions.

Currently, however, the bandwidth and QoS necessary to run such applications is expensive and is simply unavailable to the majority of broadband subscribers, let alone to the majority of Internet users, and thus applications of this sort have found little market. In most cases, IP telephony simply means a phone call and often one that is subtly inferior in voice quality to an ordinary circuit call.

At some point, IP telephony will win out over circuit voice not just because it is cheaper but because it has become better in every way. But that will not be the case in the near term.

The Mechanics of IP Voice

Any reasonably complete description of how telephone calls are implemented over IP networks would involve a lengthy description of how conventional telephone networks operate, an undertaking that could easily consume hundreds of pages. Hence I will present only the rudiments absolutely necessary for understanding how to initiate IP voice services within a broadband wireless context.

Hackers and hobbyists made the first IP telephone calls in the early 1990s, and they generally connected one computer with a sound card to another computer similarly equipped. Dial tones and ring tones were impossible in such an arrangement, and users had to agree beforehand to be online at the same time to complete a call. Such computer-to-computer IP telephony was never commercialized as a service and was also generally rejected in the enterprise.

Subsequent IP telephony platforms aimed at service providers sought to emulate many of the features and functions of the PSTN. They provided dial tones and ring tones, they permitted the use of standard area codes and telephone numbers, and they generally interconnected with the PSTN at one or more points.

Most such platforms utilized devices known as *gateways* where a voice call is transferred between the public switched telephone network (PSTN) and an IP network. Depending on the price and availability of IP backbone connections and the type of equipment at the customer premises, calls can go either way—originating as IP and being translated to TDM circuit, or originating as circuit and hopping onto an IP backbone.

A complete transit of the backbone in IP form is highly desirable because it avoids the interLATA fees customarily charged by carriers, but providing such transport has not been easy in the past for a number of reasons.

Note LATA stands for local access transport area.

Ensuring acceptably low latency and jitter is difficult in a pure layer 3–routed network, so an overlay of MPLS switching or something equivalent is needed to provide the requisite QoS. And that, as I have seen, is not always obtainable. Yet another problem arises when firewalls must be traversed. Most firewalls filter out voice traffic because it does not provide the usual acknowledgments necessary to keep the "pinholes" open in the firewall through which the voice traffic must pass, so, in practical terms, a form of spoofing is required that will emulate IP acknowledgments. Network address translation is also a problem because the ostensible destination of the traffic is a phone number, not an IP address. Finally, it is difficult to interface IP telephony systems with conventional billing and record keeping systems based on continuous minutes of usage.

All these problems have been addressed by various means in current VoIP systems, but they were not satisfactorily resolved in the prior art, hence the tardy acceptance of VoIP among the carrier community. Today where a complete end-to-end VoIP solution is required over distance, the carrier providing it will generally install devices called *session controllers* at network exchange points and sometimes on the premises of large enterprise customers. These are used to hand off IP traffic from one backbone to another or from an IP LAN to a WAN without any translation to TDM circuit.

The other key component in an IP telephony system is a *softswitch*, which is generally separate from the gateway. Softswitches perform the signaling functions required to transmit calls across the PSTN, and they presuppose that the IP traffic will be translated to a TDM circuit subsequently. In today's telecommunications environment where as yet little end-to-end IP voice traffic is occurring, a softswitch is probably a necessity, but I view the category as transitional and as likely to be obsolesced toward the end of the decade.

Increasingly, IP phone systems include subscriber phone terminals that originate the call as IP and do not require a gateway. Most such devices conform to the Session Initiation Protocol (SIP) discussed next.

Finally, if conferencing is enabled, a separate device called a *multiconferencing unit (MCU)* is generally required to connect several parties simultaneously. The central office equipment will also need to support Universal Datagram Protocol (UDP) or, preferably, RTP. SIP equipment provides inherent support for conferencing, so there is no longer a need for a completely separate platform.

A rather recent development in IP telephony is the pure peer-to-peer network, which distributes network intelligence among the subscriber nodes and requires no large infrastructure elements such as gateways or border session controllers. Skype software is a well-known example of the approach. At this point the benefits of this approach to the service provider are difficult to discern because the necessity of a service provider to manage the voice calls is all but eliminated. In the longer term, peer-to-peer IP telephony may pose a major threat to traditional voice service models, but at present the ultimate prospects of peer to peer are a matter of conjecture.

VoIP Standards

Telecommunications is an industry built on standards. Because ownership of network resources is diverse, the networks have to have some means of interoperating with one another, which necessarily entails standards. IP telephony, as it happens, has many standards. The problem is that none is a standard in the strictest sense; that is, none is universally accepted.

The first standards that were widely used in VoIP networks were UDP and the International Telecommunications Union's H.323, which is a variant of the older H.320 standard designed to facilitate videoconferencing over circuit networks. UDP is an unreliable protocol for streaming multimedia over the Internet that dispenses with acknowledgments, and it takes the place of TCP in such applications. H.323 posits end-to-end timing mechanisms and handshake arrangements for time-sensitive traffic and has provisions for converting packet traffic to circuit, and vice versa, over a gateway. H.323, which is by far the more comprehensive of the original standards, operates in a computationally intensive manner—which accounts for its waning popularity with network operators—but does little more than set up and tear down the call. It has no provisions for implementing security, managing the overall network, or determining an optimal path. Both UDP and H.323 are legacy protocols today, but both are supported in many VoIP products.

The remaining standards extant are SIP, Media Gateway Control Protocol (MGCP), and Media Gateway Control (MEGACO). Of these, only SIP is widely supported in hardware today, and it appears well on its way to becoming a real standard.

In a sense, H.323 and SIP can be placed in one grouping and MGCP and MEGACO in another. The first pair of protocols focuses on the terminal from which the IP call is launched and received, and MGCP and MEGACO focus on the gateway. SIP may be viewed as a replacement for H.323, though it is much more than that, but MEGACO definitely is intended to supplant MGCP. SIP, after a shaky start in the market a couple of years ago, is quickly establishing itself, but MEGACO is scarcely supported in actual products.

The beauty of SIP from the perspective of the operator of packet access network is that it lends itself to deployments where an end-to-end IP connection is utilized. The intelligence is largely in the terminal, but devices known as *SIP proxies* are also used to direct network traffic. For this reason, a SIP telephone call requires either a special SIP phone or else a SIP interface to which an ordinary telephone is attached and would normally reside at the customer premises, but the cost of such devices is no longer an obstacle. The aim with SIP is to avoid gateways and even softswitches per se and to utilize SIP within a larger framework of routing or MPLS switching. In this context, a telephone call is treated as just another packet stream.

Setting Up IP Voice Services

At this point, utilizing SIP-based equipment for IP voice appears to make the most sense for the broadband access provider. The price of SIP phones and interfaces has come down dramatically, making them competitive with ordinary telephone equipment, and SIP provides the surest migration path to the enhanced telephone services of the future as well as a greatly simplified network model for the IP voice provider.

Still, even with the latest SIP equipment, the network operator must not take the provisioning of IP phone services lightly. Delivering telephone services remains a highly involved undertaking requiring access to either class 4 switches or session controllers in major Internet exchanges, which in turn requires the lease of long-distance, large-capacity fiber connections as well as the negotiation of arrangements with other networks. Given these difficulties, the best course for a small independent operator may be to enter into agreement with one of the new nationwide SIP service providers, of which Vonnage is perhaps the best known. Many of these firms actively seek relationships with competitive local exchange carriers (CLECs), which relieve the latter of the headaches of dealing with long-distance network transport.

Of course, they also take a share of the profits, but then the broadband wireless operator is going to pay for transport in any event.

The other option is to purchase one's own gateway and strike a deal with a provider of circuit voice, most likely the local incumbent. Since this same entity will more than likely be competing with the wireless operator in offering broadband access, any relationship is likely to be ambiguous at best. Obviously in assigning telephone numbers to subscribers, the broadband wireless operator does need to coordinate with the incumbent telco, but it is best to avoid a situation where one is essentially reselling services from the latter, especially a situation where one is compelled to collocate equipment in the incumbent's central office.

IP Voice Service Offerings

As with other broadband offerings, the key to succeeding with IP telephony is to differentiate oneself from the competition. Since in most cases the incumbent telco offers satisfactory, highly reliable, and low-cost local phone service, achieving positive differentiation is not easy.

The strategy of most providers of broadband voice to date has been to focus on the low-cost provisioning of additional lines rather than attempting to compete in the area of basic phone service. With an IP voice system it is relatively easy to provision several additional lines with little or no installation work required, and for small businesses particularly, such a service can be very attractive. Also attractive to businesses is IP centrex, where the network service provider offers the features and functionality of a public branch exchange (PBX) from the central office.

Another unique service that a broadband access provider can offer is an IP connection between IP PBXs located in remote offices. This will enable the business to avoid any long-distance charges whatsoever and can result in considerable savings in a large organization.

Long-distance services involving end-to-end IP can also be attractive simply because they are likely to be somewhat cheaper than their circuit equivalent. Here the wireless broadband operator will have to enter into an agreement with an IP long-distance carrier.

Finally, videoconferencing is fully supported by SIP and has traditionally been a high-revenue business. Service platforms are changing so rapidly that it is difficult to characterize them, and the trend today is toward equipment that is essentially autoconfiguring and does not require operator intervention for call setup. The spread of such platforms will undoubtedly limit the role of the service provider in the future and will surely limit the fees that operators can charge, but there is still a need for large capacity, stringently managed pipes, and the network operator can definitely charge for those.

Enabling Storage Services

Storage networking is a service offering that has yet to take root among broadband wireless service operators. In the past it has involved specialized protocols such as Fibre Channel and ESCON that have not been supported by wireless broadband equipment, and the transport itself has almost always involved optical fiber. Today, however, a number of standards permit the transmission of storage traffic across IP networks, and these represent opportunities for broadband network operators utilizing 802.16 equipment.

Three of these standards are paramount. FCIP is a tunneling protocol that permits the interconnection of geographically distributed Fibre Channel storage area networks (SANs) over IP networks. iFCP is an IP-based protocol that allows for IP linkages between Fibre

Channel network elements. Finally, iSCSI is intended for connection of IP storage network elements within an IP network.

To set up storage services, a multitude of specialized switches, bridges, and gateways are required; in other words, it is not a trivial undertaking. Except in the case of a few multiservice godboxes, one cannot do storage area networking with an edge router or a metro Ethernet switch. Depending upon the particular storage application, QoS in the form of a guaranteed bit rate or latency figure may be required, and this is particularly true of operations where databases of customer profiles must continually be accessed and updated.

Storage services can also involve maintaining and operating data storage centers where storage capacity is leased to subscribers. This type of business has not done well in the past because in large part of the reluctance on the part of most businesses to entrust vital records to a third party. Also, data storage is sometimes combined with disaster recovery where the vendor mirrors vital internal networking resources at a remote facility to permit an enterprise to continue to function in the event of the physical destruction of its own headquarters. Provisioning such services is a fundamentally different business than providing local access, though, and is very capital intensive. It is not an area that a startup seeking to build a public communications infrastructure is likely to be able to enter.

Getting a Services Perspective

Setting up profitable services is not simply a matter of having the right hardware and protocol suites. What is needed above all is a correct orientation in respect to one's customer base and a correct appreciation of the competitive environment.

Value-added services are, on the one hand, a way to distinguish the network from competitors and, on the other hand, to meet the changing needs of the subscribers. A wireless broadband network operator (and any other access provider, for that matter) is as much in the business of supporting the activities of business customers and serving as a broker for communications services for residential subscribers as providing a high-speed connection. It follows that the business of the network operator is likely to evolve rapidly as new data and multimedia applications reach maturity. I emphasize that the notion that access is some basic utility such as water or power is largely outmoded and entirely insufficient to ensure the survival of the network in the years to come.

CHAPTER 8

Network Management and OSS

Previous chapters focused on service delivery extensively if not exclusively, and in them I attempted to describe systematically how a broadband wireless service network is launched, a process that extends from the initial design work and the deployment of the network elements to the selection of appropriate software platforms and networking protocols.

My focus heretofore has been on the physical layer of the network, which can never be taken for granted in any wireless operation; later chapters emphasized the higher layers of the network common to all broadband services, with service delivery forming the unifying theme in the discussion.

In the interest of clarity, I have presented each layer in relative isolation, and I have devoted relatively little attention to how the various layers communicate with one another or work together as a service-delivery mechanism. Also, I have not devoted much space to the manifold routine business operations involved in running a wireless service provider network, which, as it happens, have grown increasingly automated and conjoined with various network management operations that themselves have also become highly automated and, I may add, highly complex.

This chapter redresses that deficiency and provides the broadband wireless operator with insight into the all-important topic of operations support systems (OSS), the area of network management that will increasingly absorb the energies of the network operator after the network has been launched and built out. In a network of any size, the OSS software platforms are the intelligence that keeps the network up and running—the indispensable ingredient in any profitable service operation.

OSS: A Definition

OSS has grown to become a vast field, and the products it encompasses now occupy a multitude of subcategories. So various are the functions involved in the management of a service network that summarizing OSS in a sentence or two is difficult, but in general one can say that OSS software serves the purposes of managing the business aspects and service-delivery mechanisms within the network itself and the customer base. OSS software helps to assure that provisioning, changes of service, billing, and controlling access to the network all execute expeditiously, accurately, and positively for customers.

OSS is a catchall term for the management of connectivity and online services as commercial services, and as a software category it goes far beyond the control software used to manage network elements, though at the same time it incorporates the latter. Such control software works in tandem with other software suites within the OSS category to enable and disable services for subscribers, and in concert all of the various modules operate synergistically and transparently to link customer transactions with the network operator to actual modifications of service.

OSS in Perspective

OSS is largely a development of the last 20 years. Traditional telephone and cable television networks were not highly automated in respect to business functions, and the enterprise software used within service provider central offices seldom communicated directly with network elements to initiate, modify, or terminate service. Indeed, most of the networking elements made prior to 1990 were automatic only in respect to switching functions, and they required manual configuration for the commencement of service to subscribers. Given that the service networks were in most cases monopolies, the network operators had little need to automate the management of either the network itself or of the business functions relating to the sale of network services to the subscribers. Any costs associated with operational inefficiencies were simply passed onto the subscriber, and the many and costly human interventions attendant upon any change of service provided reassuring human contact to those same subscribers.

The rise of competition in the wake of the Bell System divestiture in 1984 coupled with the rapid increases in computing power in the years thereafter gave rise to OSS as you know it today as operators sought platforms exploiting the new generations of high-speed microprocessors, platforms that would maximize the efficiency and cost-effectiveness of their operations. As with most new developments, OSS took a few years to define itself, but the growth of the category since the early 1990s has been explosive, and even during the depressed period following the burst of the telecom financial bubble, that growth never slackened. Indeed, it increased as network operators sought to reduce staffing requirements, deliver new types of services more quickly and more responsively, and generally achieve rapid profitability in the face of declining outside investment.

It is safe to say that no network of any size can compete effectively today absent a large degree of automation. That being the case, OSS software is an investment that is every bit as necessary as radio base station radios, controllers, and subscriber terminals. Unfortunately, OSS software solutions are far less straightforward in operational terms than is standards-based wireless broadband equipment because OSS itself is governed by no similarly comprehensive standards. The situation is made worse because OSS software manufacturers tend to specialize in one or at most a few subcategories, and so the network operator is frequently obliged to follow a "best-of-breed" approach by purchasing software modules from a number of vendors to manage various aspects of the operation. The OSS industry is sufficiently mature to where specialized vendors have formed partnerships with other specialized vendors and made their respective products interoperable, but the network manager may yet face a situation where those systems that are fully interoperable may not be best of breed and where the manager may have to sacrifice functionality to achieve systemwide interoperability.

Incumbent telecommunications carriers commonly employ a large number of programmers with specific expertise in OSS to devise "hooks" that will allow favored OSS systems to communicate with one another, but a startup wireless network is in no position to emulate that practice. The network operator requires instead a comprehensive solution that is more or less fully integrated. Fortunately, some vendors will customize and modify their basic software offerings to allow the operator to achieve such integration, but the operator is still left with the task of evaluating a large number of possible software combinations and seeking expert opinion as to which combinations allow for easy integration while at the same time meeting the crucial operational needs of the network.

The Anatomy of OSS

Several well-defined subcategories of OSS software exist today along with perhaps an equal number of less distinct groupings. A number of different taxonomies emphasizing different aspects of the technology may be constructed to incorporate all of the various subcategories, but the cardinal distinctions involve the nature of the process that the software is intended to automate. In general, OSS software will either manage the network itself so as to enable various services for individual subscribers or manage the business transactions involving customer requests for service and payment for the same. A third category may have the purpose of supporting transactions with other carriers and service providers. Categories also exist for processes such as service activation and deactivation that span the two major divisions.

OSS for the Network Infrastructure

Within the first subcategory, which is network-centric, two further subcategories may be identified: network element management and network management. The first category is not necessarily the province of the independent software provider, but the second almost always is.

Network Element Management

Network element management software refers to control software for the various devices that make up the network, including routers, switches, and the radios themselves. Newcomers to telecommunications may be surprised to learn that in many cases the same basic elements made by different manufacturers may not necessarily talk to one another, and elements that are different in function and/or bear a hierarchical relationship to one another often have associated control software that functions essentially as a closed system. A synchronous optical network (SONET) add/drop multiplexer, for instance, would in its traditional guise require manual intervention to accept new customers, and even in its updated, next-generation SONET form, would not normally communicate with a metro Ethernet switch to provision a circuit for a subscriber or to increase the bandwidth of an existing connection. True, network elements exist that combine SONET and Ethernet functionality and can do these things, but they are not the norm.

The manufacturer of the particular device in question almost always provides network element management software, though that same manufacturer may not actually have developed the software in-house but instead may have gone to a software vendor specializing in this area. In addition, independent software vendors sell network element management products that may provide certain types of functionality not present in the manufacturer's own software product. Frequently what a network operator is seeking in the product of an independent is a system that will facilitate integration with other elements in the network.

Network Management

Network management software addresses such integration problems as well, though it generally does more than that. The term commonly refers to products intended to manage the network as a whole and to mediate between disparate network element management systems. The ideal is that the network management software should allow the network administrator to control any element in the network from the same computer console and indeed from the same screen menu. Every element in the network becomes completely subordinate to the ultimate service objective that the entire network must support, and when a service is ordered, the network will respond automatically to the command with appropriate actions across the chain of network elements over which the particular application must run.

Both network element management software and network management software commonly include diagnostic tools for identifying problems in the network and determining their place of origin. Many such subsystems can issue "trouble tickets" that prompt technical support staff to perform corrective actions. Incidentally, almost all OSS platforms, not limited to the previously mentioned classes, are capable of generating detailed status reports.

Network Inventory Management

A third subcategory within the overall category of infrastructure management tools involves the processes of *discovery and inventory management.* For a network management system to operate effectively, it must utilize an accurate database that lists all the network's physical resources, including hardwired sections of the network as well as network elements, and that provides a full accounting of where and to whom available bandwidth, telephone numbers, and Internet Protocol (IP) addresses are allocated. In a small operation with a few dozen customers, the network operator may well be able to memorize such information, but in a fully evolved metropolitan network with hundreds or thousands of customers, that is clearly impossible. Thus, inventory management software becomes a virtual necessity.

Network Planning

A final subcategory that stands a little apart from the others is *network-planning software.* I have already covered software used for siting base stations and determining coverage, but overall network-planning software goes beyond the mere positioning of equipment and attempts to determine equipment needs in respect to every layer of the network such that the operator can achieve service-delivery objectives in the most cost-effective manner. Unfortunately, most network-planning software developed to date has not been designed with the specific needs of the wireless broadband operator in mind but instead has been aimed at the wireline contingent.

OSS for Customer Relations and Transactions

The second major division of OSS has for its objective the automation of the service desk. OSS will not entirely eliminate the need for telephone agents to answer queries and deal with complaints, but what it will do is handle most back-office functions. Where in traditional telephone networks invoices would have to be made out for each change of service and work order issued so that the changes of service would be enabled on central office equipment, provisioning software largely exempts these processes from human intervention. In fact, Web-based provisioning for credit card transactions is entirely possible today.

Provisioning

A large subcategory of OSS customer care products is concerned with *provisioning*, that is, setting up services for existing accounts as well as signing on new customers and providing them with services. Provisioning software can completely eliminate the need for a customer representative for routine transactions and can save the network operator substantial amounts of money for that reason. Normally such software customers will be directed to a secure Web site where, after authenticating themselves, they can request new services or alter existing ones. The provisioning software will then communicate back to the inventory management software to determine whether the resources are available to support the new service and will then signal the network management software to command the necessary operations to take place in the network elements to enable the service. Additional communications will be made to the billing and mediation software to correct a customer's account to reflect the changes. In the case of new customers, a credit card may be requested as well as an address. Obviously, with a new customer, the process cannot be completely automated because a subscriber will still have to be provided with a terminal, but in many instances that can be mailed to a customer's residence and self-installed rather than involving a truck roll and a visit by a technician.

Some provisioning software is capable of throttling bandwidth to users who attempt to command a disproportionate share of network resources, and such capabilities are valuable and desirable. Bandwidth is still a scarce resource in wireless networks, and low-value customers should not be permitted to monopolize it.

The actual implementation of service is sometimes defined as a separate process and is known as *service activation.* Normally service activation is accomplished through the network management and element management modules, not through a specialized software solution.

Billing, Mediation, and Service-Level Agreements

Billing and mediation software forms a large subcategory of OSS. Such software will usually be tied in with the provisioning, network element, and network management software to register changes in service and to track subscriber usage if the service plan is based on anything other than flat-rate billing.

Here a word is in order on billing plans and service-level agreements: The usual pattern in the formative period of broadband access services was to charge subscribers a flat rate for service and allow customers as many transmissions as they desired. Such "all-you-can-eat" service plans still prevail in the residential broadband market, though they are declining among business-oriented broadband offerings in favor of what are known as *tiered services.*

Tiered services are just what the name implies—stratified offerings where higher service fees command higher speeds and additional capabilities. Such plans allow the network operator to tailor services to individual users and make certain that the subscriber is charged only for the bandwidth and network resources actually utilized, and it also allows the operator to regulate "bandwidth hogs" such as individuals running peer-to-peer video and music file-sharing sites or call centers, as well as online gamers who may spend hours at a time in bandwidth-intensive transactions.

Increasingly, tiered service plans are accompanied by service-level agreements (SLAs), which bind the service provider to meet certain quality of service metrics stipulated in the agreement. Such metrics may include minimum throughput, latency, jitter, and bit error rate—anything under the control of the network operator and for which the subscriber is willing to pay. Such agreements usually provide for penalties to be imposed upon the service provider for failing to meet the terms of the agreements—generally a reduction in the service fee. Incidentally, it is best to be scrupulous in such matters and to agree only to terms that can be met in normal circumstances. To be wildly overoptimistic regarding what the network can deliver is to verge upon fraud.

Assurance

Assurance OSS software refers to the management process of determining that changes in services and the corresponding changes in billing have in fact been carried out and, specifically, that stipulated service levels are being met.

CRM

Customer relations management (CRM) software as a product category encompasses a vast array of software offerings, by no means all of which are designed around the needs of telecommunications service providers.

The term most commonly refers to software used in call centers by sales agents and telemarketers, though it can also include software for facilitating online transactions involving sales, modifications of services, and service requests.

CRM will normally tie in with billing software and will also communicate with a central database of customer profiles and subscription information.

Data Mining Software

Data mining refers to a specialized type of analytical software that looks for patterns and relationships within the information contained in a comprehensive database, and it forms a class unto itself that is not directly connected with the larger customer relations division of OSS inasmuch as it does not involve flow-through procedures involving other software modules.

In the case of telecommunications networks, data mining could be used for a number of purposes. The network operator may want to examine the demographic skew of certain service offerings as indicated by the existing customer base and then adjust marketing strategy accordingly. If, for instance, enterprises of a certain size or in a certain type of business tended to be frequent users of conferencing services, the network operator may want to construct a marketing campaign for acquainting similar types of subscribers with conferencing services on the theory that they will be apt to purchase such services. Or, if a certain demographic grouping tends to abuse flat-rate unlimited services, one may want to manage the network so as to throttle bandwidth to such users and put provisions for doing so in place before the commencement of service. Or, to cite yet another example, if a group defined by certain attributes recorded in the database is subject to unusually heavy churn (telecom jargon for customer turnover), one may decide to avoid active solicitation of individuals within that group or make special efforts to determine the source of the churn.

Knowledge in the broadband access business is power, and data mining can provide network operators with the same kind of in-depth information of user preferences amassed by traditional market research companies. Even so, data mining has not been extensively used by independent operators and has remained largely the province of large incumbents (though it is by no means universal even there). Like all statistical techniques, data mining requires a reasonable sample to produce accurate results, so in a small network it is of dubious utility.

OSS Software Integration

As indicated earlier, the best-of-breed approach prevailing in the service provider community in respect to OSS has led to major software integration problems—problems for which, unfortunately, no easy solution is in sight. Industry groups such as the Institute of Electrical and Electronic Engineers (IEEE) and the International Telecommunication Union (ITU) have devoted tremendous efforts toward standardizing hardware and have achieved commendable results, but such standardization deals only with the lowest layers of the network. It is certainly an important first step toward providing an integrated, ready-to-deploy infrastructure for the broadband wireless network operator, but it provides no real basis for managing the business aspects of the network.

The relative immaturity OSS software *sui generis*, and especially the difficulty in integrating various software modules, could plausibly be the most difficult problem facing the wireless broadband operator today. And it is a problem for which only partial out-of-the-box solutions are available. But difficult as such integration problems are, network operators must address them. They simply have no choice.

That being the case, the following sections cover the options.

Protocols for Software Integration

In existence today are protocols and standards providing for the exchange of information among network elements and associated software control systems (that is, network element management software modules and network management software modules). Two such protocols are used more than any others: Simple Network Management Protocol (SNMP) and Common Object Request Broker Architecture (CORBA). Both are generally concerned with supporting machine-to-machine communication among network elements, but their approach is quite different, and CORBA represents a far more sophisticated and far-reaching technology.

SNMP originated in the enterprise world and is designed to collect management information from devices on the network. It presupposes a centralized network management software system. SNMP uses a request-and-response process to obtain information from the participating devices and consumes little bandwidth in transmitting such information, which makes it robust. It will provide such information to the centralized control console as network topology, traffic patterns, and diagnostics, and it is capable of disconnecting nodes as well as permitting centralized management of said nodes. What it is not designed to do is directly control devices in the network or to support automatic flow-through processes where one event triggers another. An example of the latter is flow-through provisioning when a subscriber orders new services, and subsequently the billing, provisioning, network management, network discovery, and network element modules all automatically respond in a sequenced fashion.

SNMP is a well-proven protocol, and nothing is wrong with it, but its intended use is in internal local area networks (LANs) where issues such as billing and SLAs are irrelevant. It is simply not designed to support commercial service offerings. The majority of wireless broadband products made today support SNMP, and that is as it should be, but such support should not be considered a complete solution to the operator's software OSS integration problems.

CORBA, the other major protocol for achieving network integration, reflects a revolution in software development that took place in the late 1980s and early 1990s, namely, the rise of object-oriented programming. That revolution was entirely successful, and today its effects are absolutely pervasive not only in telecommunications but in countless other software applications.

Object-oriented programming is basically a modular approach to creating applications where standard tools are included with the data to be manipulated, and the combination thereof is referred to as an *object.* Object technology can easily integrate many types of information, and it allows different applications to communicate with one another readily. Objects themselves can readily be imported and reused across applications, and programmers and developers can assemble new applications out of predefined code sequences. Lengthy books have been written on the subject of object-oriented programming, and their length has been justified by the power of the concept and the way in which it has simplified and rationalized the programming process. Given the huge number of Web-based applications developed in the late 1990s, mostly through object-oriented techniques, one can say that object-oriented programming arrived just in the nick of time. The Internet today would be a different and far less interesting place had the revolution not occurred.

Insofar as network elements and associated OSS systems contain objects, and most of them do today, CORBA provides a way for objects to communicate with other objects, which in turn provides a degree of interoperability that permits flow-through procedures and a high degree of network automation. For this reason, CORBA compliance is quite commonplace though far from universal in current network elements.

In addition to SNMP and CORBA, a number of other protocols aimed at easing interoperability in the service provider network are extant in the marketplace. These include Common Management Information Protocol (CMIP), Common Management Information Service (CSMIS), and a number of Java initiatives endorsed by the Java Community Process organization shepherded by Sun Microsystems. The Java offerings consist of Java 2 Enterprise Edition (J2EE), Operational Support System/Java (OSS/J), and New Generation OSS (NGOSS).

CMIP and CMIS are closely related, and, regrettably, neither has found much acceptance among manufacturers. At this point they do not appear to represent the future. CMIP can to a certain extent be regarded as an outgrowth, embellishment, and replacement for SNMP, but it is a far more complex and powerful protocol, permitting the execution of various tasks as well as the accumulation of information from network elements. But, because it is based on the obsolescent Open Systems Interconnection (OSI) model promoted by International Standardization Organization (ISO), it holds increasingly little appeal for either OSS platform developers or equipment designers.

Java-based systems, for whatever reasons, have not developed a large following either, and it is safe to say that CORBA is as close to being a de facto standard for system integration within public networks as any protocol in use today.

OSS Application Interfaces

Many OSS software developers follow a markedly different approach to OSS software integration by working toward standardization of common application programming interfaces (APIs) rather than comprehensive interoperation standards. Such APIs permit software to exchange information and pass through commands in regard to specific operations, but they do not form the basis of a common unified control plane.

In the face of such unsatisfactory standards-based solutions to integration, most network operators strive to form a primary relationship with one OSS vendor that offers a number of modules and form other relationships with a sufficient number of specialized vendors such that something approaching a single turnkey system may be imposed upon the wireless network. Because no two diversified OSS vendors offer complete solutions but rather solutions that are incomplete in various particulars, and because the needs of individual network operators diverge depending on the size and scope of the network and the types of services offered, I hesitate to make specific recommendations. Some of the OSS companies that offer software suites capable of handling more than one aspect of OSS are NetCracker, Ai Metrix, Granite Systems, Cramer Systems, Eftia, Convergys, Cisco Systems, MetaSolv Software, Telcordia Technologies, and Syndesis. Unfortunately, none of these companies makes specialized products aimed at wireless broadband service providers. NextNet, a manufacturer of broadband wireless radios and base station equipment, makes such a system, but it is intended for use with NextNet's own equipment.

Wireless broadband, it must be said, lags behind other access technologies in deployment, so software developers see little possibility of achieving large sales volumes. And large amounts of investment are not flowing into broadband wireless networks, so this makes them even less attractive to the OSS software community.

Summation: The Well-Managed Network

Managing the network for profitability remains the biggest challenge facing broadband wireless operators. Installing and operating the physical plant is fairly straightforward; the integration of the various kinds of business software necessary to make it all function smoothly with little routine human intervention is not. Network operators are advised to select OSS and security platforms early in the planning process, rather than waiting until the network build is completed. Attempting to operate a network manually without automating the various business functions initially is almost certainly to begin operating at a deficit, with the prospect of deepening that deficit with every day that automation is forestalled.

CHAPTER 9

Network Security

Security is not part of operations support systems (OSS) proper, but it is an integral part of managing a broadband access network. Indeed, security, in terms of the integrity of the network's own infrastructure, the safety of its customers, and the nation itself, is becoming increasingly important in network operations today and cannot be considered optional at this point.

Network security is a broad subject covering a number of areas. The most significant of those areas have to do with securing the network elements themselves. These encompass securing vital databases, including those concerned with customer records, network inventories, transactions with other service providers and carriers, and general business financial records; preventing unauthorized access onto the network and, in particular, preventing entry into customer virtual private networks (VPNs) or customer local area networks (LANs); preventing or limiting denial-of-service (DoS) attacks; and finally meeting CALEA reporting requirements imposed by the federal government in addition to other related regulatory mandates.

Note CALEA is the Communications Assistance for Law Enforcement Act.

As well as securing network elements, software platforms, and customer and business databases, one has to consider securing a whole other area of security dealing with facilities management. Because of the prominence accorded to hackers in the news media and in trade publications, the emphasis today is on data security, primarily on protecting network elements and the information they store from malicious code, but such concerns should not blind the network operator to the entire range of security concerns. Safeguarding the network from hacks performed over the Internet is certainly a worthy objective but is far from the only area upon which the network operator should focus.

Security Policies

Every business today, including a public service network, requires a security policy that will be rigorously monitored. Also, all aspects of security administration should follow that policy. The policy should be holistic, including threats involving physical intrusions into the facilities and not just remote attacks over the Internet.

Specifically, the network operator should have a policy for controlling human access into the central office. The central office itself should either be guarded or be equipped with a secure locking system that will keep out unauthorized individuals. If the central office is not staffed at all times, then it should be equipped with a surveillance and intrusion-detection system sufficient to thwart entry until humans can respond. Increasingly, municipal governments use broadband wireless networks for their own communication needs, and the governments must be reassured that the network hub is not wide open to attack.

All vital records should be backed up in a secure facility and transmitted to that facility over a secure virtual private network (VPN) if transmission over public networks is involved. The VPN should make use of encryption and not just tunneling. In many cases, storage will take place over a private internal network and stored data will reside in a storage data array within the central office. Of course, the network can be presumed to be secure if it is dedicated to storage and is not accessible from the outside. If it is accessible, then the storage network must be protected with a firewall just as is the case with any other network.

All network elements performing vital functions should be replicated such that a reserve unit can be immediately pressed into action in the event of a failure. Many carrier-class network elements have built-in redundancy where every aspect of the system is replicated internally.

Most network elements made today utilize card and cage construction, and, when that is the case, individual cards should be hot-swappable so that the entire device need not be shut down to replace a card. The aim of the network operator must always be to minimize downtime.

Secure Electrical Systems

Another part of security and good network management is to make certain that high-quality electrical power will be available at all times even in the event of a power outage. This involves several distinct measures.

The AC power provided by many electrical utilities is often remarkably inconstant, exhibiting long-term and short-term voltage sags as well as overvoltage conditions and occasional spikes where voltage levels may exceed the standard voltage by many multiples. The AC may also be troubled by the presence of *harmonics*, distortions in the AC waveform that can disrupt the functioning of many kinds of electrical or electronic components if sufficiently severe. All these conditions are undesirable, and some may be catastrophic, and the network operator must guard against them by appropriately selecting power conditioning and power backup equipment.

Power conditioning devices take a number of forms.

Passive systems consist of high-frequency filters (of limited usefulness because they cannot raise or lower voltage or eliminate harmonics), constant voltage transformers, and switched tap autoformers. Constant voltage transformers and switched tap autoformers are devices that will maintain constant voltage within certain values, say, 5 percent over and under the nominal value. While both are essentially passive in their operation, switched tap autoformers contain logic circuits and relays that select among output taps on the autoformer coil to compensate for changes in input voltage. Constant voltage transformers operate on a different principle; the transformer core is partially saturated at the nominal line voltage and will grow more or less saturated as the input voltage goes up and down, which in turn will cause compensatory changes in the output voltage. Constant voltage transformers are marginally more reliable than switched tap autoformers, but they also tend to be more expensive.

A better solution is an uninterruptible power supply (UPS), which is always on and is series-connected between the wall power and the devices being powered. Carrier-grade UPSs are examples of active power-conditioning devices. Such a unit usually consists of an isolation transformer interfacing with the wall current, an array of rectifiers that transform the alternating current into direct current, a bank of storage batteries to hold the DC charge, and a set of high-frequency solid state switches that converts the DC back into 50- or 60-cycle AC. Higher-quality UPSs have provisions for maintaining a steady output voltage regardless of input voltage and also perform *power factor correction*, a process that eliminates harmonics in the output of the UPS caused by reactive electrical loads associated with switching power supplies in computers, with cathode-ray tube flyback circuits, and with most electrical motors.

The better UPSs produce a smooth sine wave output, and lower-quality units produce a coarse, stair-step waveform. Since stair-step waves are rich in noise and distortion components, they are undesirable.

Some UPSs recently introduced into the marketplace utilize fuel cells either in lieu of batteries or to supplement them. Although generalizations must be made with caution in this area, it is safe to say that most types of fuel cells produce several times the amount of energy per kilogram as most types of secondary batteries, though the lowest energy density fuel cells and the highest energy density batteries almost overlap in this regard. Fuel cells are currently very expensive, minimally $5,000 per kilowatt, but prices may begin to decline in the near future. Fuel cell backup power may begin to become the norm toward the end of the decade. High-velocity flywheel generators are also beginning to appear in some central offices for providing highly reliable, though relatively short-term, backup power.

The central office facilities may also use backup generators. In most cases, these will use ordinary diesel or gasoline reciprocating engines, but a growing trend is to use devices called *microturbines*, which usually run on natural gas. Microturbines are made by such firms as Capstone, Allied Signal, and Ingersoll-Rand, and they are derived from the turbine designs used in jet aircraft. Currently microturbines are much more expensive than diesel engines.

Diesel generators should be routinely tested because often diesel engines will fail to start when they have not been operated recently.

In all cases, backup power must come on automatically and instantaneously in the event of a power failure. The subscriber should experience no interruption of service whatsoever.

Cyberwarfare

Whole books have been written on the subject of software-based network sabotage, and they will continue to be written simply because hacker tools and utilities are continuing to evolve. In this section I can suggest only the rudiments of a policy for dealing with such attacks.

Operators of public networks have a greater obligation to secure their operations against hacks and cybersabotage than do ordinary businesses because the public depends on the services they provide. Above all, the network operator is selling reliability, and system downtime attributable to hacks is intolerable.

The problem in meeting security requirements on the part of a network operator is that network security is a full-time job. Network security officers in large enterprises must spend a considerable portion of their waking hours lurking around hacker Web sites simply to keep abreast of developments, and of course they must also familiarize themselves with the torrent of security bulletins pouring out of various monitoring organizations. Security administrators cannot afford to fall behind in such matters because their systems are immediately at risk if they do so. Obviously, an ordinary information technology (IT) manager entrusted with the routine administration of the network who tries to do security in idle moments—which scarcely exist in that position in any case—is not going to be successful.

A large, mature network will probably find it wise to hire a security administrator, but a small startup generally cannot afford to do so. The only solution then becomes the retention of a reputable security firm—in other words, the outsourcing of security.

This is not necessarily a bad idea. Specialists in the field such as Computer Security Associates are thoroughly up on the latest hacker strategies and will undertake aggressive network defense, including legal action against attackers. Such services are not inexpensive, but simply hoping attacks will not occur and doing nothing may represent a false economy.

It is a good idea to have such a network security company perform a security audit on the network infrastructure from time to time as well as provide routine updates on security software and response to individual problems. The audit should encompass not only the OSS and the vital databases but also the facilities themselves, including the central office and base stations.

A word about overall security policy and securing the network against software attacks: Network operations staff should as a matter of policy not be permitted to download files either from the Internet or from privately recorded discs onto computers utilized in network management. It is also a good idea to attach individual firewalls to such computers to prevent the former practice. In any case, the policy should be explicitly stated and rigorously enforced. *Trojan horses* are a favorite weapon of hackers for gaining access to well-secured networks. Network operators should also be alert to the possibility of internal sabotage by disgruntled employees. Many security organizations have suggested that the majority of computer crimes are inside jobs. Finally, visitors should not be allowed free access to vital network elements or left unsupervised in their presence, and this applies to authorized maintenance personnel. Institutional paranoia is a good adaptive response for any organization running a vital services network.

Attacks and Counterattacks

Hackers have a variety of motives, and their ploys tend to reflect that fact. Some regard network intrusion as a harmless sport and do little or no damage after they have achieved access. Others regard sabotage itself as a sport and intrude in order to destroy. Others engage in industrial espionage, seeking to steal information and sell it for a profit. Still others are hired assassins seeking to wreck a network at the behest of a competitor. Yet another group intrudes primarily to steal software for redistribution. And a surprisingly large number of hackers seek to enter a network to use it as a launching platform for further attacks, thus disguising the ultimate point of origin of such attacks.

In the case of public networks, hackers may attempt entry not to attack or compromise the access network itself but to breech an enterprise network attached to the public network. Or they may want to eavesdrop on private transmissions either out of voyeuristic motivations or for financial gain.

The arsenal of tools used by experienced hackers today is enormous, much too large to be discussed in this chapter. Unfortunately, such tools are readily available as freeware at hacker Web sites, of which there are hundreds if not thousands. And the ready availability of such tools has had extremely unfortunate consequences. In the 1970s and 1980s, hackers tended to be young computer professionals, and because the knowledge base required at the time was so extensive, not too many of them existed. Today any computer-savvy adolescents with a yen to hack can easily equip themselves with the weapons to do so without understanding the mechanisms by which they operate. We are also seeing an increase in the activity of cybercriminal gangs who commit computer crimes for profit. Many of these organizations operate in Eastern Europe.

Fortunately, security software has kept pace with the democratization of hacking, and the security professional now has a large arsenal available. And while the number of products on the market is considerable, the basic approaches they embody are not numerous, and the network operator should be able to easily comprehend them.

Cybersecurity Technology

Previous chapters have already covered firewalls. Firewalls are the first line of defense for the security administrator, but they should not be considered complete security solutions in and of themselves. Closely related to firewalls and sometimes included in the category are *proxy servers*, which are devices where information requested from a database is actually launched onto the network or where applications are executed on client software remote from the main server. Proxy servers protect vital information and programs from direct access by outside parties, and they limit damage to nonvital facilities in the face of a network attack. In other words, they serve as buffers.

Diagnostic software detects the presence of malicious code and unusual activity within the network. Antivirus scans form a subcategory within this grouping, though they are not the only such products to which such nomenclature applies. Diagnostic software may be roughly divided into two primary divisions: software used in security audits to determine the overall vulnerability of the network and software used routinely to detect anomalies. In both cases, the developer must continually update the software for it to remain effective.

Some such software has the ability not only to determine the nature of an attack or intrusion but to find its point of origin—in other words, to follow the hacker back to a home base even across multiple networks. Such software must also be updated more or less continuously, since skilled hackers are always finding new ways to disguise their activities and identities.

Security professionals often use *encryption software* to render vital data unreadable to hackers. Modern encryption methods are highly effective, and encoded material can only be decrypted by intruders who have access to massively parallel computing systems running for weeks at a time. Encryption techniques today use *rounds*, which are successive reencryptions that can number in the millions and make the encrypted data seem more and more random and meaningless. Essentially, there is no way to decrypt such messages by clever insights. Instead the intruder has to try out all possible codes one by one with a specialized decryption program. With enough computing speed, almost any machine code can be cracked, but such speed is not available to a lone hacker with a Pentium processor.

Business records, customer profiles, and billing information should be routinely encrypted and should never be presented where they can be intercepted in decrypted form. Encryption is also advisable in VPNs.

To sound a cautionary note, if grid computing services (see Chapter 3) become generally available in the future, then hackers will have a formidable weapon for decrypting formerly secure information, and at that point the encryption industry will have to come up with new approaches. But currently, encryption remains a powerful preventive tool for the security administrator.

Finally, within the arsenal of defensive procedures, some software engines are designed not only to detect malicious code but also to prevent its effects by restoring network data to its state just prior to the detection of suspicious activity. Such software is a fairly new development, and it may not be entirely effective against all conceivable attacks.

Authentication is sometimes considered a part of security and sometimes just a part of routine network operations. In a wireless network authentication, the process by which network users demonstrate that they are who they purport to be is especially important because the physical layer of the network is essentially open. Authentication today is normally performed in specialized servers, most of which now run Radius software.

Safeguarding Network Elements from Hijacking and Malicious Code: Best Practices

Securing network elements is, of course, vital to the integrity of the operation. While direct attacks on equipment operating systems intended to disable networks for lengthy intervals are not at all commonplace, intrusions into management systems have occurred in the past and undoubtedly will occur in the future. Obviously, they should be prevented at all costs. Unfortunately, many of today's network elements are more vulnerable than the telco "big iron" of the past. Telephone circuit switches and asynchronous transfer mode (ATM) switches generally utilized some variant of Unix as an operating system and involved extremely arcane code that few hackers ever mastered. Indeed, most of the people who successfully hacked into telephone central offices were experienced individuals working in telecommunications. In contrast, many network devices manufactured today use open or commonly understood platforms such as Linux, Windows NT, or Java. They may incorporate some type of software firewall to thwart intruders, but they are not inherently difficult to understand or manipulate. And because wireless transmissions can be physically intercepted with great ease, there is little physical layer security possible in the network, and the network operator must remain largely dependent on specialized security software.

The danger is compounded because most equipment today is designed to permit remote management by an authorized network administrator from a supposedly secure Web site. Obviously, that greatly eases the job of network administrators, enabling them to respond to problems in the network anywhere and at any time without having to visit the central office. However, if the administrator can access the OSS suite, then so can a hacker—if that individual can get past whatever security measures are in place. Accordingly, the network operator must make certain that there are no "trap doors" permitting entry into the management system that bypass authentication measures.

Denial-of-Service Attacks: A Special Case

Denial-of-service (DoS), or *flooding* attacks, have been used for many years by hackers. They are launched by capturing a large number of terminals and using them to transmit meaningless messages that flood the public network affected beyond its carrying capacity.

DoS attacks are different from viruses and worms. They do not introduce malicious code into network elements or management software, and they are transitory in their effects. And yet they can shut down a network effectively for hours. Recently several vendors have developed software to detect the presence of DoS attacks and to prevent network elements from continuing to pass the bogus transmissions that characterize such attacks.

CALEA and Other Regulatory Burdens

Communications Assistance for Law Enforcement Act (CALEA) is a complex piece of federal legislation that expands law enforcement's authority to conduct electronic surveillance, including surveillance over public networks. It has been recently reinforced by provisions contained within the USA PATRIOT Act, authorizing a broad range of surveillance and interdiction techniques conducted over public networks.

■ **Note** The USA PATRIOT Act is the Uniting and Strengthening America by Providing Appropriate Tools Required to Intercept and Obstruct Terrorism Act.

To date, the provisions of CALEA have primarily been applied to incumbent telcos, if for no other reason other than the act was conceived in the area of circuit-switched networks and is somewhat difficult to apply to pure packet networks such as those characterizing broadband wireless deployments. Nevertheless, if the current political climate and preoccupation with terrorism persist, you may expect that the efforts of law enforcement to monitor the public networks will intensify and will place increasing burdens and liabilities on the network operators.

Already the federal government has used such programs as Carnivore to inspect traffic in a wholesale manner to determine the presence of suspicious activity, and legal barriers to fairly indiscriminate monitoring have been considerably relaxed. Where this will lead is anyone's guess, and the legal and societal implications of such practices are best left to those with specific expertise in constitutional law and political processes. However, that such activities will increasingly involve the service provider in an uncompensated role may be predicted with a good degree of confidence, provided, that is, that counterterrorism remains at the heart of governmental policy.

Advising the network operator on how to cope with such exigencies is difficult because of an almost utter lack of prior experience on which to draw. In the recent past, various authoritarian regimes have attempted to restrict and monitor the Internet activities of their subject populations, with mostly indifferent results. The governmental firewall around the island of Singapore, for instance, is relatively transparent and ineffective. Nevertheless, network monitoring on the part of law enforcement agencies is likely to become more extensive in the future, though whether it will become more welcome depends on external political circumstances. In any event, the necessity of abetting the practice puts the network operator in a nearly impossible position. On the one hand, one wants to be perceived as concerned about national security matters and ready to do one's part to check further attacks upon lives and property here or abroad. On the other hand, one's subscribers expect their privacy to be protected. And if such broad surveillance powers are abused, as they have been in certain instances in the past, then the position of the service provider becomes more difficult still.

At the present moment, government-monitoring requirements do not impose an onerous burden on wireless broadband operators. But the possibility that they eventually will is not entirely remote. Unfortunately, a technical manual can offer the network operator little guidance in these matters. Ultimately one's own beliefs regarding privacy and openness and, conversely, on one's own obligations to the state, will play a critical role in the decisions one will make regarding supporting monitoring activities. Some conundrums really are conundrums. Some choices remain difficult regardless of how much professional expertise one commands.

Network Security: Business Overhead or Another Profit Center

Security costs money. That is why businesses and even governments are reluctant to allocate funds for security unless they perceive a dire necessity for doing so. Additionally, security is a matter of degree, never an absolute, and determining appropriate security expenditures becomes more difficult still.

One way to make budgeting choices involving security less painful is to explore ways in which security can be made into a profit center. Encryption devices can provide secure VPNs as well as encrypt internal records, for instance. They can also protect content intended for sale over the network such as entertainment programming. Network operators can also offer monitoring and other value-added security service to enterprise customers, or they can offer security software for resale along with other business applications.

Security, in the last analysis, is simply a means for controlling access to various parts of the network. And controlling access lies at the essence of running a profitable network.

Index

B

D

E

F

V

W